# FEW AND FAR BETWEEN

# FEW AND FAR BETWEEN

*On the Trail of Britain's Rarest Animals*

*Charlie Elder*

B L O O M S B U R Y

LONDON • NEW DELHI • NEW YORK • SYDNEY

*For my parents,*
*Andrew and Penny*

Bloomsbury Publishing Plc

50 Bedford Square
London
WC1B 3DP
UK

1385 Broadway
New York
NY 10018
USA

www.bloomsbury.com

BLOOMSBURY and the Diana logo are trademarks of Bloomsbury Publishing Plc

First published 2015



Author photographs taken on his camera are with thanks to Sammy (great auk egg), David (common skate), Becca (great crested newt) and Ben (vendace). Spiny seahorse, as seen by author, photograph taken by Neil Garrick-Maidment.



British Library Cataloguing-in-Publication Data
A catalogue record for this book is available from the British Library.

ISBN: HB: 978-1-4729-0518-5
PB: 978-1-4729-0519-2
ePub: 978-1-4729-0517-8



Typeset by Deanta Global Publishing Services, Chennai, India

Printed and bound in Great Britain by CPI Group (UK) Ltd, Croydon CR0 4YY

To find out more about our authors and books visit www.bloomsbury.com.
Here you will find extracts, author interviews, details of forthcoming events and the option to sign up for our newsletters.

# Contents

# One

I am sitting alone beside the sea at dusk on a remote island in the Hebrides, staring at a blob of peanut butter. I've been watching and waiting for well over an hour, and I'm cold and tired. A strong westerly wind that has raced across the Atlantic, trailing its fingers in the ocean, is flinging spray across my back and trying to elbow me out of its way, while the wet pebbles beneath me have squeezed all feeling from my calves. This is an exposed place to spend time in any weather: a causeway of round grey stones, smooth as peppermints, heaped up by the waves on both sides and linking two outcrops of land. At the far end, a towering shoulder of basalt shelters the nests of puffins and razorbills beyond, while in front of me vertical slabs of rock topped by tough grass and sea pink rise from the tideline. It is beneath these cliffs, tucked in among clumps of stranded seaweed, that I have placed a spoonful of peanut butter before retreating to a suitable vantage point close by.

I can't deny that my behaviour is slightly bizarre, and should probably have come with an advisory notice to passers-by: 'Warning, this beach contains nuts.' Not that you get passers-by in this far-flung location. The isolated Shiant Isles, situated off north-west Scotland, aren't the kind of place where you just happen to find yourself unless shipwrecked – and if you came across a lone figure sitting in the gloom, hood pulled up over his head, camera clutched in one hand, gaze fixed on a smear of Sun-Pat, you would be forgiven for turning around and swimming back to your sinking vessel. Even more so if you knew what I was looking for. Over two days I have journeyed hundreds of miles, by plane, coach, car and boat, to get here in the hope of encountering a small mammal most people would happily travel a long distance to avoid, a rodent that in its heyday was responsible for wiping out half the population of Britain yet is now a national rarity. And despite the discomfort, the cold and fatigue, being bullied by the wind and deafened by crashing waves nearby, I'm filled with excitement, rooted to the spot and scanning the rocks in the fading light for movement, for a glimpse of dark eyes, a hairless tail . . .

This is one of the final destinations on a quest that has taken me the length and breadth of Britain in search of our rarest and most endangered animals. I have visited wetlands, heathlands, woodlands and uplands, endured extremes of weather, been bitten, stung, pecked and shat on (from a great height), had a go at pond dipping, fishing, snorkelling and moth trapping, joined shark, bird and bat surveys, and figured out how to get a good night's sleep in a Ford Focus. I am now close to completing the challenge I set myself a year ago, and ready to return to family life and months of unbroken work, having used up all my holiday entitlement. Yet even though I have come so far and seen so much, I would gladly start out all over again. Far from having satisfied an appetite for wildlife, I am hungrier than ever for new encounters.

In the meantime, trusting to good luck and an offering of sweet peanut gloop, I am hoping to spot just one more scarce species, and as I shift position, lifting my head to stretch the stiffness from my shoulders, I notice out of the corner of my eye something stirring among the boulders. This is one of those magic moments on the trail of rare animals that I have come to love, when months of anticipation, weeks of planning, days of travel and hours of waiting come down to a split second, an intake of breath, a heartbeat of hope coursing through the veins. Could this be it? Could this be it?

# Two

A smooth snake basks in a patch of warm sunlight by the heather. Natterjack toads grate the air above the dunes with their rhythmic croaking. Deep within a nest of dried honeysuckle a dormouse sleeps curled in a ball. Young pine martens tumble in play on the forest floor. Ah, the great indoors!

Once again I found myself slumped in front of the telly, cradling a glass of wine, watching breathless presenters enjoying exhilarating views of British wildlife. I had managed to outwit my two teenage daughters and seize control of the TV remote, and was adopting the airline steward approach to parenting: 'Always attend to yourself before your children.'

'Haven't you seen all this before?' my eldest daughter sighed.

'Perhaps,' I admitted.

'Can't you just go out in the garden and look at stuff there instead?' her sister suggested as they shuffled out of the room.

'I'm tired.'

'Lame!' she called out from the stairs.

There was a time, before developing an allergy to adult enthusiasm, when they might have watched wildlife programmes with me, but not anymore. They were too busy cultivating their own interests to pretend to share mine, and, growing up in a village on Dartmoor, it was perhaps natural that they took our flora and fauna for granted. I had spent my younger years in London and still grasped at outdoor experiences, wrapping my arms around them, clinging on desperately, inhaling deeply, sucking them up as if they might escape me: 'Look, a sparrowhawk!' 'What a view!' 'Come and see this frog!' Over the years there had been notable high points when I congratulated myself that a little excitement had rubbed off on them – like when my daughter, aged 16, actually noticed a bright red bird in Central Park on a trip to New York and demonstrated commendable identification skills as she pointed it out, remarking in a deadpan tone: 'Look dad. A bird.' And there had been low points, such as the time I took my youngest daughter on a ski-centre toboggan ride in Plymouth, which involved racing down a long metal half-pipe on wheels. During her first run, a mouse jumped into the smooth-sided track and was unable to escape, and with no steering she couldn't avoid hitting it. As if that wasn't traumatic enough, I had bought a multiple-run ticket, so, knowing I am not one to waste money, she had dutifully climbed aboard her toboggan again for her remaining goes – flattening the dead mouse five times in all. We are told our children should connect with nature, but perhaps that was taking things too far.

I was a little younger than my daughters when I first became interested in wildlife, and it wasn't a sudden revelation, as can happen with boys taken to a football match or on catching their first fish. I can't recall being given a pair of binoculars for Christmas, exciting as that might have been, or visiting a menswear shop for an anorak fitting. Instead my fascination developed gradually on long annual family holidays that combined destinations of unspoilt beauty in Scotland with nothing much to do. When I wasn't passing the time outside trying to break and burn stuff with my brother, I was exploring on my own, wandering in the fields and along the shore, attempting to put names to birds. For a city lad these unbounded spaces were both liberating and unsettling. Wilderness wasn't a place where you could hide – unlike London it actually noticed you. Nature was watching, and I felt it around me, inviting me to look, to listen, to understand.

My growing curiosity came at a time of increasing public concern about the plight of our planet. We had seen the first photograph of Earth taken from space, read about the effects on wildlife of pesticides and pollution, watched news footage of dying seabirds washed up in slicks of oil and images of flattened swathes of rainforest, and worried about the impact of our spiralling population. The threat of nuclear annihilation and the reality of species extinctions provided the cheery backdrop to my teenage life. Growing up during the Thatcher years in Muswell Hill – a left-leaning area of north London where you needed five tokens from *The Guardian* to qualify for residency – I was used to protest marches as fairly regular days out, but fears for the environment felt fairly abstract given the fact that I spent my time stuck in a classroom, a bedroom and a bus. It was only on visits to the countryside – those holidays in Scotland and family visits to the Lake District, Dorset coast, Cambridgeshire and Gloucestershire – that I got a glimpse of a world worth protecting.

And then I closed the blinds for fifteen years. I ended up working in newspaper offices in cities and towns, and left the trees to hug themselves. Fires, crimes, health scares, housing plans, road accidents, charity walks, elections and funding cuts filled my days. I was far too busy hammering out wedges of text about what mattered to other people to think much about what might matter to me, and it was only when I joined a paper in the West Country, and moved with my wife and young daughters to Devon, that I emerged blinking into the sunlight, as pale as a health-food shop worker, and realised what had been missing from my life: wide horizons, night skies peppered with stars, muddy country lanes free of road markings, the smell of livestock, fat hedgerows . . . I was also aware of how much there was to learn about the wildlife around me. What was that high-pitched squeaking in the long grass at the end of the garden? The flourish of song from the trees? That small butterfly with folded green wings? Those birds circling in close formation above the hillside opposite?

With the help of guidebooks I was gradually able to tease out a little sense from the tangle of sights and sounds, to understand a few of the whats, wheres, whens and whys that governed the diversity on our doorstep. The list of species stuck on the kitchen pinboard grew year by year, and there were always surprises that set the heart racing: a water rail beside the garden stream one winter; an otter chanced upon in a nearby river; two snow buntings on the moor; heath fritillary butterflies in a woodland glade. Unable to contain my excitement at every new sighting, I told anyone prepared to listen, ploughing on regardless through their polite nods and drifting gazes. And it was birds that really inspired me, so much so that I ended up travelling around Britain in my spare time on the trail of our most threatened species, writing a book that ensured family and friends already wearied by my ornithological anecdotes had the additional pleasure of having to read them.

I had by now reached my mid-forties, a time when things can go badly wrong, and I knew I needed to get a grip when I found myself wondering whether I ought to know more about wines and thinking out loud about learning a foreign language. After my travels in search of endangered birds, I had stalled a little. The vegetable patch had become overgrown, the dog walks were getting shorter, the chickens and ducks had been found new homes, and the beehives were left unchecked. I loosened my grip on spare time and work started to steal back the minutes and then the hours. My daughters were right to mock me for slumping in front of nature programmes. I had probably spent more time watching wildlife on television over the preceding year or two than actually going out and seeing it for myself. Sure, I had spotted plenty of birds in passing, but they were only part of the picture. Why on earth hadn't I seen a smooth snake, a natterjack toad, a dormouse or a pine marten? I certainly wanted to, but they weren't the kind of common or widespread species you just bumped into. You had to make the effort.

It was time to get on the road once more. And I had a plan, of sorts. I switched the TV off, went into the dining room and declared: 'Right, I've finally decided, I'm leaving.'

My wife knew what I was talking about, given I had been chewing over ideas for a while, and replied without looking up from the computer: 'Well, make sure you take the recycling on your way out.'

During my previous travels I had become fascinated by the scarcer species I came across, those rarities clinging on in small numbers and teetering on the brink of extinction in the UK. Whether thriving or not elsewhere in the world, they are part of the unique variety of life on these islands, and I found myself increasingly drawn to them and concerned about their fate. Scarcity bestows a certain quality, an undeniable allure, which, with stars of the conservation cause, can approach celebrity status. One can feel privileged to be in the presence of rare animals – the honour of a brief audience with a secretive Scottish wildcat in the snow in 2010 was a case in point, remaining among my most treasured memories. Yet the thrill of encounters is tempered by a range of deeper emotions. Witnessing living creatures in the knowledge that their kind are struggling for survival is a genuinely moving experience. With diminished numbers we begin to see them as individuals, fragile and special, like each of us, one of a kind. They embody a celebration of diversity in the natural world and the tragedy of loss that pricks at our conscience, and their vulnerability whispers truths about the ephemeral nature of existence and our own temporary grip on life. They also reflect two sides of our attitude to the environment: a callous disregard that puts ourselves before the welfare of the planet, and a caring and compassionate desire to protect wildlife and put wrongs right. That certain species are in danger of vanishing says one thing. That they are still here can say another.

My plan was to go in search of Britain's rarest and most endangered species – not just those with feathers, but also fur, fins and scales, four, six and eight legs. Of course I wasn't heading out of the door for good, as my wife was well aware, just devoting weekends and work holidays over the course of a year to the pursuit of a selection of scarcities scattered like precious stones across our land and surrounding seas. It was a daunting challenge, and harder still knowing that I wouldn't be at home much over the months to come, but I felt sure that my family would forgive me – after all, what teenagers wouldn't be bursting with pride at the thought of their father heading off to track down a rare newt or moth? I was sure they could hardly wait to tell friends at their local comprehensive school.

The difficulty was deciding which animals to see. So much of what we have is becoming scarcer by the day. A staggering sixty per cent of species have declined since I was born in the mid-1960s, with hundreds close to disappearing and several dozen gone altogether. Most of our native bird species are decreasing, three-quarters of our butterflies and two-thirds of moths are in trouble, sea fish stocks have plummeted, many of our bumblebees and bats are losing ground, a third of dragonflies are under threat and wildflowers have been disappearing from counties at the rate of one species every year. Our skies are emptier, our summers quieter than they once were, yet the gradual pace of change means that it can be hard to discern absences. The present drags the baseline with it and becomes a new starting point – the way things are. Imagine, then, 44 million more breeding birds sharing our lives. That is what we have lost since England won the World Cup in 1966 (along with a few penalty shoot-outs). And if you were to rewind a few generations to the 1800s, you would find our feathered friends so plentiful that millions were being culled as pests or for food. Hedgerows, woods, fields and gardens must have been buzzing with birds, and the air filled with movement and song. Little wonder everyone at that time wore hats.

But not all declining species are scarce, and not all scarce species are in decline. Some uncommon animals existing close to the precipice of extinction may be relatively stable, while more everyday species that are haemorrhaging numbers, for one reason or another, present a real cause for concern. Rarity is not the same thing as threat; it is a slippery concept that ignores how well species are faring and finds its footholds in maths and geography – abundance and range. And the definition distorts depending on the area you are mapping, so that a species might be considered plentiful locally, rare nationally and numerous globally, all at the same time.

For the most part, though, the picture is pixelated through a lack of data. On a worldwide scale we know next to nothing about the status of the vast majority of species, and from what we have gleaned, one surprising irony emerges: rarity is relatively commonplace. Abundant organisms make up the minority of species. They may dominate vast swathes of the planet by sheer weight of numbers, particularly where the living is easy, but alongside them and on the fringes, a multitude of species has evolved to exploit myriad niches. Of course, predators such as tigers and eagles, which are at the top of the food pyramid, will always be restricted in number, but it goes further than that, with rarities of all kinds sprinkled liberally across tropical mountain tops and isolated islands, rainforests, reefs and scrublands and almost everywhere between. While we are undoubtedly making species rare, much of the variety of life on earth is born rare.

In Britain, those species we regard as scarce tend to be more widespread or common elsewhere, because our wildlife is generally shared with our neighbours in Europe. But let's forget them, swatting away swallowtail butterflies with their baguettes. What's rare in the UK is what matters most to me. I began by asking conservation organisations and experts for their suggestions, and read as much as I could about species populations and threatened biodiversity. Despite some tough choices, I was eventually able to piece together a list of manageable proportions – not escapees, introductions and anomalies, but scarce and significant native wildlife. And what a tantalising mix of weird and wonderful creatures awaited me, from a monstrous predator of the deep, a scaly lizard-eating constrictor and a fearsome meadow dweller with a talent for minor surgery, to an adorable nocturnal furball, a bird with a song straight out of *The Lord of the Rings* and an elusive insect 'terrorist'. . . twenty-five animals in all, distributed across the country: five birds, five mammals, five reptiles or amphibians, five fish and five invertebrates, each species with a story to tell about life on the edge, each ready to dive for cover as soon as I came into view.

And I began with a bit of jelly and an egg.

It was the summer of 1972, and Oxford zoology graduate Richard Ivell had risen early, eager to press on with his research. He stepped into his backyard to sort through several buckets of mud and brine collected the day before. These thick, stinking sections of sediment dug from Widewater Lagoon in West Sussex provided raw material for his master's thesis on the ecology of brackish environments. Sandwiched between land and open sea, lagoons are vulnerable and yet hostile habitats, where fluctuating temperature and salinity test life forms to the limit. Landlocked Widewater, a shallow stretch of water running for nearly a mile next to the A259 near Worthing, provided the perfect place to study. Bordered by the back gardens of houses to the north and beach huts to the south, the narrow basin is kept topped up by rainfall and by the sea that percolates through its shingle banks during high tides. Various kinds of prawns and cockles live within its confines, and some of these creatures would have been scooped up in the samples Richard had collected just a couple of metres out from the seaward edge. The mud had settled overnight within the buckets, each surface a dark brown disc submerged under a lens of clear salt water. In the quiet, life had stirred.

Forty years later, the memory is still vivid, as Professor Ivell told me on the phone from the animal biology institute in Germany where he now works. Peering over the rim of one of the containers, he couldn't believe what he saw. 'In the stillness, small anemones had emerged and their lightly banded tentacles were spread out flat against the sediment. The thing I remember is how beautiful they were, and how fragile, like tiny flowers in a desert. It was quite amazing to see.'

Unable to identify the small burrowing species, he showed a colleague at Oxford University, Richard Manuel. If the catchphrase among Anthozoa taxonomists is 'know thine anemone', then Dick was the person to ask. An expert in such marine life, he realised that these specimens, no more than a couple of centimetres long, buff coloured with twelve transparent tentacles, the outer nine marked with cream bands, were new to science.

'He was very excited by the find and he named the anemone after me, which was extremely flattering,' Professor Ivell said.

A complicated mix of factors, from topography and geology to climate and the hand of man, have created a remarkable array of habitats in Britain, with a wealth of creatures to match. But we have very little flora and fauna we can exclusively call our own. The last ice age pretty much wiped the slate clean, as ice deep enough to bury Ben Nevis covered the land as far south as Finchley Road tube station in north London, and the eight thousand years since meltwater created the English Channel and cut us off hasn't provided much time for new species to evolve. While a few subspecies are going their own way, particularly on our remote islands, full-blown endemic species are very precious indeed. Ivell's sea anemone was an important discovery because it was found only in Britain and nowhere else on earth.

Which also makes it one of our most important losses.

One can only speculate as to the causes, but changes in the lagoon environment meant that within a decade it was gone, declared extinct, and it has not been seen since, despite several searches.

'It is a very timid animal and I only discovered it by chance,' Professor Ivell said. 'I wouldn't exclude the possibility of it being found somewhere else, but knowing the fragility of these habitats it also wouldn't surprise me if it has simply disappeared, as there is nowhere suitable nearby where it could live.'

The demise of Ivell's sea anemone is not exactly widely known, but as Britain's only reported endemic animal loss of recent years it is significant. Before I set off in search of our scarcest and most threatened animals, I felt I needed to pay my respects to this tiny jelly organism, the delicate rarity we failed to safeguard, which only a handful of scientists and laboratory technicians ever saw alive.

I drove on a bright March morning to the Oxford University Museum of Natural History, where I met up with assistant curator in the zoology department Dr Sammy de Grave, who, like Professor Ivell, has a species named after him – a colourful shrimp pictured on his computer mouse mat: *Salmoneus degravei*. In fact, as one of a tiny coterie of world shrimp experts, he could have plenty more bearing his moniker. On trips to far-flung locations he has discovered more than a hundred new species, and has dozens of specimens awaiting formal description.

Following our phone conversations, Sammy had kindly arranged a special delivery from the museum vaults. There on his desk, resting on a brown velvet base in a small wooden-framed glass box, was a large egg. The label, yellowed with age, read 'Egg of the Great Auk – *Alca impennis*. Bequeathed by Sir Walter C. Trevelyan. Bart. 1879.' One of our most famous extinctions, the great auk was a large flightless seabird that looked like a cross between a razorbill and a penguin and was once common off our northern coasts and across the Atlantic. Clumsy on land and easy to catch for food, feathers, fuel and fish bait, it was mercilessly hunted to extinction by the middle of the nineteenth century. The elongated white oval shell before me was one of only a small number of eggs that still exist in private and museum collections. Bigger than a goose egg, with holes chipped in either end where its contents had been drained, it was marked with spots and smudges of dark chocolate-brown that congregated in soft swirls around its broad base – a unique pattern that not only enabled parent birds to recognise it as their own, but also added to its appeal among Victorian egg collectors.

It is not known where the egg came from, but today this hollow promise of a life never lived, snatched like a battlefield souvenir amid remorseless slaughter, is a poignant reminder of the permanence of extinction. Having read a great deal about the tragic extermination of such an impressive bird, how the last great auk in Britain was stoned to death on St Kilda by superstitious locals fearing it was a witch, while the final known pair were strangled by fishermen on a remote Icelandic island a few years later in 1844, I felt extremely fortunate to have been able to see a great auk egg up close. And hold it. And not drop it.

But it was Ivell's sea anemone that I had really come to see, and Sammy led me along winding corridors to an invertebrate storeroom tucked away behind the main building. Unlocking the door, we stepped inside, the air sharp with the tang of chemicals. Long fluorescent strip lights flickered into life on the low ceiling, revealing rows and rows of jars that had been shut away in darkness. Tall, squat, wide, thin, the glass containers were lined up on shelves that ran along the walls and in rows across the centre of the small, windowless room. All manner of creatures were suspended within them in preserving solution – crabs clenched like bony fists, jellyfish discs, tangles of segmented worms, sponges, sea slugs, an octopus with tentacles pressed like ribbons of bubble wrap against the sides. Many had lost their pigment, which gave them a ghostly embryonic quality, and some of the pale older specimens from the 1800s had turned the liquid that surrounded them the colour of straw. On one shelf, a selection of crustaceans and insects were marked simply: 'Mr C Darwin'.

'From the voyage of HMS *Beagle*,' Sammy told me.

In a corner cabinet stood several jars with red tape around their tops, which contained holotypes, the original single specimens from which species were first described, and there among them was one bearing the scientific name *Edwardsia ivelli* – Ivell's sea anemone. A number of test tubes were visible inside, floating in the clear ethanol solution, one labelled 'holotype' and the others 'paratype', denoting further examples. Lifting the jar to the light, Sammy looked for a while in silence. 'Hmm,' he said eventually. 'I suppose it must be in there somewhere.'

'What, nothing there?' I strained for a glimpse of an organism floating within the test tubes. It was hard with labels in the way, but they did seem empty. I had been expecting to fail with sightings of rare living animals, but a dead one, sealed in a jar in a locked room, felt like a very bad start indeed. Could it have been stolen? Swiped by some black-market trader in pickled invertebrates working for a ruthless crime lord and trafficked overseas inside Anthozoa 'mules' before being peddled in a shady backstreet deal? Had the anemone underworld beaten me to it?

Sammy, who had himself searched in vain for Ivell's sea anemone at Widewater Lagoon, wasn't going to be beaten again. He suggested we take the jar back to his room to examine the contents properly.

Once there, he opened the lid and emptied the test tube marked 'holotype' into a Petri dish. I was relieved to see a twist of something solid in the centre of the liquid, no longer or thicker than a nail clipping from a little finger. Was that it? I knew Ivell's sea anemone was smaller than the fat red anemones you get in rock pools, but I hadn't bargained on it being quite so tiny. Little wonder it took until the 1970s to discover one. It resembled a snippet of white string, slightly frayed at the head end where the tentacles were visible – not much to look at. But through a lens it was transformed into something of wonder, from the foot that would have been buried in the mud and its transparent tubular body to the crown of tentacles draped around its mouth. Delicate, simple and intriguing, I couldn't resist looking again and again.

'You are possibly the only person who has examined this find under a microscope in forty years,' Sammy told me.

Blimey. I kept quiet about the fact that the last time I had looked at anything through a microscope was during school biology classes. Instead, trying to look like I knew what I was doing, I twiddled a couple of knobs on the side, and then couldn't get it back into focus.

As extinctions go, Ivell's sea anemone seems like a miniscule loss, a species we can surely live without. Why on earth should its disappearance matter?

'Just because it's small doesn't mean that it wouldn't have an important place in an ecosystem,' Sammy said. 'It clearly had a function where it lived, and its loss may have had an impact on other species.'

Perhaps even ourselves. Compounds in sea-anemone venom have shown potential in the treatment of everything from multiple sclerosis and arthritis to cardiac problems and obesity. Though regardless of whether it had some value to humans, this otherworldly life form, strange enough to have arrived on a passing meteorite, was surely worth saving for its own sake. The truth is, we've shown ourselves to be remarkably good at taking our wildlife for granted. It is estimated that 480 plants and animals have gone from England alone during the last two centuries, including such eye-catching species as the large tortoiseshell butterfly and breeding Kentish plovers, alongside which a creature as primitive and alien as a sea anemone is always going to struggle in the popularity stakes. Across the globe, species are disappearing for good more than a thousand times faster than the expected natural rate of loss, with almost a quarter of animals assessed considered to be threatened with extinction. Imagine then the fuss if the multi-billion-dollar Mars rover *Curiosity*, currently exploring the surface of the red planet, came across a salty puddle with a few little anemones at the bottom. You can bet mission control wouldn't let out a collective bored sigh and instruct it to trundle on past.

'Life!'

It's as precious here as there.

I had one more museum to visit before I started my search for Britain's living rarities, only this one was a little closer to home. Plymouth City Museum and Art Gallery has among its possessions a national treasure that has won admirers wherever it has been exhibited, from its first public outing at the Royal Academy in 1885 to the present day. Painted by one of the founding members of the Newlyn school, *A Fish Sale on a Cornish Beach* is one of my favourite works of art, and it tells a story, not just of the locals engaged in trade and gossip at the water's edge, but of the richness of our seas.

Stanhope Alexander Forbes was tutored in London and the studios of Paris before settling in west Cornwall, inspired by the quality of natural light that has drawn so many artists to the West Country coast. Here he set about trying to capture everyday lives on canvas, following the French vogue for working *en plein air*. In the summer of 1884, he set up his easel close to the old quay in Newlyn and began working on what was to become his masterpiece. Five foot wide by four foot high, Forbes's study of fishermen selling their catch took a year to complete – exhausting no doubt both for himself and for the local models paid to stand for hours exposed to the elements.

Near the top of the picture, luggers with tarred hulls and darkly tanned sails can be seen anchored in the bay, while fish are rowed to shore. An auctioneer, or 'jowster', as he would have been known, stands at the edge of the receding tide and rings a bell, gathering a crowd of women in heavy aprons, who inspect the haul at his feet. In the foreground, one of the fishermen, holding a hand line and a few mackerel, pauses to talk to two women leaning on a small rowing boat and seated on a wicker fish basket. Theirs is a compelling scene, an unhurried encounter over a selection of fish, the perspective so inclusive that you feel you could take a step forward and interrupt – it is well worth visiting Plymouth City Museum if ever the painting is on show, simply to try to overhear what they are saying. And between them, strewn amid dull reflections on the wet sand, is the second story: the impressive size and selection of species to have been landed by the inshore boats: halibut, bass, cod, John Dory and, spread belly side up, common skate.

Today Newlyn is still a major fishing port, but the view captured by Forbes at the end of the nineteenth century, when fish were plentiful and cheap, has altered a great deal. Two piers now cross the stretch of sea framed in the painting, with much of the sand dredged for their construction, and gone are the beach sales and the luggers. Gone, too, is the common skate. This bottom-hugging giant, slow to grow and slow to breed, was once widespread around pretty much all our coasts, including the English Channel, but has been hoovered up in seabed trawls so extensively over the decades that it is now largely confined to areas off north-west Scotland, the Shetlands and the south-west of Ireland and is anything but common. The prefix hangs from its name like a lead weight. It is thought that there may be only a few thousand of these leviathans left in British waters, perhaps a few hundred in some key areas, and in a desperate attempt to stem the decline, commercial fishermen have been told they are no longer allowed to target or land them – closely related ray species making up our takeaway helpings of 'skate and chips' instead. The International Union for the Conservation of Nature (IUCN), the global organisation that assesses the status of species, is so concerned about the plight of the common skate that it has classified the fish as critically endangered – one step away from the exit door for an animal that has been around for 150 million years. Not even the snow leopard, tiger or panda are assessed as being at that level of threat.

The problem is that the common skate is relatively easy to catch – in fact, it is hard to avoid for any fisherman dragging a net across the ocean floor. Even young skate are too large to slip through the mesh, spooked from the seabed into the mouth of the advancing tackle. Given that this long-lived predator, which starts breeding when it is at least ten years old, tends to remain faithful to favoured sites, it doesn't take long to clear them from an area for good.

They may not be the most attractive creatures lurking at the bottom of the sea – and given what you find down there, that's saying something – but they are certainly impressive beasts, well equipped for hunting in the cold and murky depths. Pressed flat against the sediment, their olive-brown bodies would be hard for passing prey to spot, and their large eyes, keen sense of smell and skin peppered with pores sensitive to electromagnetic fields ensure they miss little going on around them. They are also surprisingly fast, enabling them to add such quick movers as mackerel to a diet of basement-level fish and crustaceans. And just in case things get hairy, they sport a powerful tail bristling with sharp thorns for protection – not that fully grown skate, capable of downing dogfish in a single gulp, have much to fear.

Fish behaviourist Dr Viki Wearmouth, who is based at the Marine Biological Association just a short stroll from Plymouth City Museum, knows all about those tail thorns. She was releasing a skate as part of her monitoring work when its tail slipped between her hands and the blades cut into her finger. A painful encounter, but she hasn't lost her love of these giants.

'They're a fascinating species, so immense, with many unanswered questions about how they live, and are much more active than is assumed, coming off the seabed to hunt,' she told me. 'But because they live out of sight in the sea, it can seem like they don't matter as much as threatened land animals, and people are less inclined to do something about their decline. It's not like you're going to get campaigns to save them, like you do for the panda.'

Unfortunately you can't sit on a rock five hundred feet down under the sea with a clipboard, trying to work out how far skate roam and what they need to survive. Instead, researchers such as Viki hitch rides on sport-fishing boats in order to tag specimens with electronic data loggers to get some idea of their movements. Handled carefully, common skate have proven remarkably robust when it comes to catch-and-release, some being reeled in time and again over the years with seemingly no ill effects. The era when anglers returned to port and posed with their catch hanging from a winch before the carcass was dumped off the end of the harbour is long gone. Today it's barbless hooks, a quick photo, a check for tags and a few measurements before skate are returned to the water. The fact that conservationists are welcomed on board at all says a lot for how times have changed, and Viki believes sport angling that follows good practice shouldn't have any detrimental effect on our skate populations.

If I wanted to see a live common skate, she recommended I call Scottish boat skipper Ronnie Campbell, who has been running charter trips for over twenty years, fishing for these monsters in the deep glacier-gouged channels between Oban and Mull – one of their last British strongholds. And I was in luck: he had a group booked for Easter Monday, a couple of weeks away, and room for me on board. All I had to do was get to there on time and pray for calm weather.

When you live in the south-west of England, most places seem like a long way away, and north-west Scotland is not exactly round the corner. I wasn't sure I had the stamina for a ten-hour drive to Oban, and the train fare was so expensive I would have had to sell my car to cover the cost, so instead I opted to fly, offsetting the carbon footprint by avoiding doing the ironing for a week. Recognising that a traveller and his money are easily parted, I brought with me all the food and drink I could carry, which was a mistake because everything liquid was promptly confiscated by airline security. I hadn't reckoned on being subjected to the same scrutiny as international travellers, and had to say goodbye to my cartons of apple juice and terrorist-scale supply of black cherry yoghurt.

Despite it being the final day in March, spring still seemed a long way off. This had been the chilliest start to a year for half a century. Snow capped the mountains that rose on either side of the road north of Glasgow Airport, and Oban felt shrink-wrapped in still, cold air, which squeezed the warmth from layers of clothing and pressed its thumb over chimneys so that they choked on their own smoke. After checking in to a B&B and parking my hire car nearby, I pulled on a hat and gloves and wandered down to the harbour, where I was relieved to find the sea was smooth. The conditions, as forecast, meant the trip should go ahead. It struck me as surprising, though, that while the historic town and its scenic coastal setting are popular with tourists, so many visitors were out in the cold admiring the view, strung out along the Esplanade, scanning the bay with binoculars and even telescopes.

That evening, after sorting out what I needed to bring the next day, I struggled to get to sleep. My journey had begun in earnest, and I was filled with anxiety and excitement. When it comes to fishing there is, quite literally, a thin line between success and frustration, between the times one never forgets and the hours one never gets back, and Oban was a long way to come to fail. From the few sea-fishing trips I had been on in the past, I was as prepared as I could be for disappointment, the ebbing away of hope as hours pass without a bite, buoyant banter wanes, answers run dry and juicy baits lowered into seas teeming with possibilities become miserable little scraps in an empty ocean. This time I could do nothing to improve the chances of striking lucky – I wouldn't be fishing, just offering two additional pairs of crossed fingers and relying on the skills of the anglers I was accompanying.

I was awake before my alarm the next morning and, after a breakfast of porridge and Kwells, walked down to the North Pier by 8am, noting once again a few people along the seafront gazing out to sea through binoculars. Ronnie welcomed me aboard his 35ft *Laura Dawn II*, and I was introduced to the fishing party, a friendly group from Shropshire and Powys: John, sporting an impressive white beard; timber mill worker Paul; a teacher called David; and another David, a fishing-book publisher accompanied by his young son, William. I was relieved to learn that they were far from boat-fishing novices – I was in the company of anglers as opposed to 'danglers'. Three had landed common skate before and were back to do battle once more with the biggest rod-and-line prize in our seas after the porbeagle shark.

Ronnie took up his place in the cabin and the rest of us stood on the open rear deck, leaning against the rails as we motored out of Oban, past the solid tower block of St Columba's Cathedral at the shoreline, round the northern tip of Kerrera island opposite and out towards the Sound of Mull.

'So did you see the whale then?' one of the group asked casually.

'Sorry, the *whale*?'

'The one that's been hanging round Oban Bay, the sperm whale.'

'Sure, the sperm whale,' I laughed, wondering where the joke was leading.

But they weren't pulling my leg. A twenty-tonne male, seemingly in good health, had been dodging the ferries and plying the sheltered waters close to shore for a couple of days. I could hardly believe it. And while all those people who had puzzled me on the quayside, scrutinising the waves with binoculars and telescopes, were sharing breathtaking views, I had missed out. It was a real blow, and I was annoyed at myself for not being more inquisitive, for writing off the onlookers as boat spotters – even though I wasn't sure such a hobby existed. Then again, how could I have imagined that they might be enjoying such an unlikely sighting as a sperm whale – that deep-diving, squid-eating, giant-headed Moby Dick of a whale that only occasionally visits our waters and seldom close enough to see from land.

When compiling my list of target species I had scribbled in the margin the vague term 'cetacean' – the group encompassing whales, dolphins and porpoises – and added a question mark after it. While a number of such marine mammals are regulars around our coast, sightings of rarities are typically chance encounters, almost impossible to plan. You could spend a lot of time sitting on headlands or ferries hoping for a fleeting glimpse of a distant fin breaking the surface, and even then it would be largely pot luck what happened to be passing by. It seemed like too much of a lottery, given that I aimed to seek out specific scarce species, and so I had left the question unanswered. In any case, I reckoned I had used up my share of whale-spotting good fortune in the recent past. On a two-day trip to Shetland I had witnessed a pod of killer whales hunting seals in the cove below my B&B, so missing a sperm whale, the once-in-a-lifetime answer to my rare cetacean conundrum, was perhaps the way it was meant to be. And there was no way I was going to pluck up the courage to politely ask everyone on board whether they minded terribly if we put skate fishing on hold for a few hours, turned the boat around and spent the morning whale watching instead. 'Um, excuse me but I was just wondering . . .' Nope. That was not going to happen.

It didn't take long before we had reached a wide stretch of water near the island of Mull and Ronnie dropped anchor and cut the engine. The clouds had cleared and in the distance the silver-backed bulk of Ben Nevis was visible. Around us, undulating land plunged into the sea, the steep gradients sloping down to an underwater valley floor 450 feet below, where dogfish patrolled and, somewhere, common skate lay flush against the bottom as if pressed flat by the sheer weight of water above them. Rods were pulled from their stands, fitted into metal holsters around the boat, and hooks baited with whole mackerel or squid before being lowered into the grey-blue. Then it was just a case of waiting, backs turned to the chilly breeze, hands warming around mugs of tea, listening out for heavy reels turning and watching the rod tips bending rhythmically in the gentle swell as lines tensed against their two-pound weights on the bottom far beneath us.

'This is where they tend to be, in the deep hollows. And people come from all over Britain, even Scandinavia, to fish for them,' Ronnie told me as he tidied away the remaining bait.

'It's the size that's so impressive,' said John, who had been skate fishing for a number of years. 'Even an average skate is big compared with other sea fish, though they're very placid and lovely looking out of water, with eyes like a spaniel.'

'But a big skate is a killer to land – really hard work to get to the surface,' Ronnie continued, adding with a laugh: 'Some people say the best day's fishing is when you *don't* catch one.'

After an hour or more I was wondering whether that might be the case. The baits were untouched and the playful jibes had begun.

'Know any good spots skipper?' David, the teacher, grinned.

'Any chance you could take us there?' the other David added.

Ronnie, as calm as you like, wasn't going anywhere, and lobbed back an old favourite: 'You should have been here yesterday . . . '

The conversation was interrupted by the boat radio crackling to life with a message from the coastguard: '*Navigation warning for Oban Bay. There is a large live whale in the bay and boats are advised to keep clear.*'

In case I'd forgotten.

And then talk returned to tackle and tactics, the sharing of experiences and expertise, until a low rumbling sound cut through the chatter and in a split second the atmosphere changed. Line had begun to spool off Paul's reel. His bait was on the move, steadily dragging monofilament off the turning drum down into the water. It was a mesmerising sight – like some primitive form of communication, telling us we had made contact with a life form from another world.

I was surprised at how calm everyone remained. Instead of leaping to grab the rod, Paul left it untouched and busied himself fitting a back-support harness and a protective pad around his waist to prevent the butt digging into him. Then, carefully removing the rod from its holster, he waited for the word from Ronnie and at his signal clicked the reel into gear and lifted into a colossal weight at the other end. The rod bent like a cartoon frown and he cried out: 'Fish on!' How he could tell it was a fish and not a sunken supertanker I'm not sure, because it didn't appear to budge an inch. Getting a big skate to shift off the bottom was the hardest part, I was told. And that certainly looked to be the case. The combined flexing of muscles and rod seemed to make little headway, until eventually Paul managed to raise whatever was at the other end ever so slightly, gaining a precious few turns of the reel as the boat dipped in the swell. At any moment it seemed like the line, as taut as piano wire, would simply snap, and it was incredible that such a thin length of tapering carbon fibre, bowed over the rail, could take the tension – I hardly could. But bit by bit Paul managed to edge his catch upwards. It was the equivalent of hoisting a garage door the height of Birmingham's BT Tower, and his grimacing face, flushed with the effort, showed what he was going through.

After twenty minutes or so, Ronnie removed a panel at the rear of the boat in readiness and we peered over the side, following the line into the darkness, until a diamond shape appeared. As it neared, I could make out its pointed head, lighter spotting on the back, and flashes of white beneath as wide pectoral 'wings' beat against the water. Then, with a final turn on the reel, it broke the surface and was swiftly manoeuvred onto the deck like a thick wedge of carpet. The hook was removed, revealing its underside and strangely human-like lips concealing mouthparts tough enough to chew up lobsters, before everyone stood back to admire it, spread out more than five feet wide on the boards, a common skate, prehistoric and yet majestic.

John was right about those spaniel eyes – they were unlike any I had seen in a fish before: it appeared to be watching, taking things in. And the deep green irises surrounding dark pupils told another story. They are considered a clue to identity that suggests specimens like this caught off west Scotland could be far rarer than first thought. For not only is the skate on thin ice, but it may also be the victim of a classification error. At one time, a larger northern variety of common skate, like this one, and a paler-eyed southern cousin were recognised, until they were lumped together under the single name in the 1920s. Now genetic studies have revealed that they should have remained separate. The common skate is most probably two distinct species: two causes for concern, splitting our population estimates at a single stroke.

Taxonomy tussles aside, it was an awe-inspiring sight. After Paul had a few photos taken with his first-ever skate, I was able to get a closer look and ran my hand over its back, glossy-looking but rough as sandpaper, to the base of a tail that was missing. It was likely this fish had been caught before, pulled free from a trawl net with a loop of twine around the tail and released with the tourniquet still in place, which had acted over time like one of those rubber bands used to dock lambs.

It was measured, the weight calculated at 185 pounds, and after a final moment of appreciation returned to the sea, beating downwards from the light until it was gone. I wished it well.

Three lost bites followed, and Ronnie decided to keep spirits up by predicting another fish at exactly 3.45pm. He was a minute out. A common skate took one of the baits at 3.46pm and fishing-book publisher David, harnessed to the rod, reeled it in over a punishing half an hour. Although slightly smaller, it was a second large female, in perfect condition with tail intact, and it had a thin yellow tag in one wing as part of a sea angling monitoring scheme, which, as it later transpired, indicated that it had been caught three years earlier. David gave it a kiss on the head and it was swiftly back in the water. No one loves skate more than those who seek to catch them. It is one of those paradoxes that the conservation movement has to embrace among a multitude of motivations for protecting threatened animals, which range along a scale from sentiment to self-interest. A primitive desire to outwit, to catch and even to kill, runs deep, and whether bagging big game or butterflies or pond dipping as a youngster, it can help forge a special bond with nature that, among the more enlightened at least, may translate into support for the safeguarding of spaces and species.

Ronnie is passionate about the welfare of skate and angling techniques that ensure they are returned unharmed. 'If every fish caught on this boat had been killed as a result, it would have had a real impact on their numbers,' he said. 'I wouldn't be in business now.' A simple philosophy, and one the wider fishing industry has been forced to heed.

Seeing two common skate and sharing the experience with such an enthusiastic group had been a wonderful experience. But the day had one more treat to offer up. Heading back into Oban Bay, Ronnie called out from the wheel: 'There it is!'

Breaking the surface less than a hundred feet away with a blow of spray was the sperm whale, the top of its huge head and grey back clearly visible as it turned in front of the harbour. An astonishing sight. And we couldn't help but draw closer as our paths almost crossed, enabling us to see its forty-foot length revealed in sections as it swam – a curve of ridged backbone, a glimpse of tail flukes – leaving us guessing at the bulk hidden beneath. This was no plankton-eater. The sperm whale is a toothed colossus, the largest predator on earth, and just about the only animal in our seas capable of dining out on a fully grown common skate. I was extremely lucky to have seen it – a few days later it swam back out to sea.

My travels had got off to an amazing start. This rare animal-spotting business is a doddle, I told myself.

Maybe, just maybe, I was getting a little carried away.

# Three

We've all been there: lying awake at night, unable to sleep, fretting about the cold and the weather forecast for the next day, wondering whether we will get to see a newt.

Or maybe I'm in the minority on this one.

Either way, the icy start to 2013 continued and the first week in April might as well have been the first in February. The light snow that had sent children in the village where I live into a delirium, as if the world had been dusted in cocaine, was gone, but the temperature hardly nudged above zero. The open moor was solid underfoot, the grass a mad calligraphy of frozen tangles. Ice covered the puddles (why do we have an urge to put a heel through it?) and the sky, still waiting for swallows and songbirds to arrive, was blank as a drum skin. At the bus stop, teenagers staged shivering protests against sensible clothing. This was not a good time to go newt spotting.

I'll be honest, I'm no expert when it comes to newts. I would struggle to hold my own at a dinner party if the conversation turned to herpetology. And you can only change the subject so many times. But the good news is that when it comes to Britain's native newts, there aren't exactly many to learn. Take up an interest in moths and you have nine hundred of the larger varieties to get to grips with. Our ground beetles number around 350, regular birds top 250, and there are nearly sixty breeding butterflies to memorise. Newts? Well, just three. And that is half the total number of amphibians that reside here. Situated at the northern limit of distribution ranges and cut off by the sea, we have embarrassingly few native species compared with our European neighbours: in short, the common frog, common toad, natterjack toad, palmate newt, smooth newt and great crested newt. Hop across the Channel and you could find double that on the other side, and plenty more as you travelled further south. (Note to self: never invite a French herpetologist to dinner.)

None of our newts could be thought of as very rare, but the most threatened is the biggest of the three, the impressive great crested newt, and we can at least congratulate ourselves on being a European stronghold for this declining species. Measuring roughly a handspan in length, black and warty with a blaze of orange or yellow beneath, it resembles a fearsome dragon in miniature, especially the male, with its jagged crest. I was keen to include it on my target list because I had never seen one in the wild before.

If I'm truthful, I hadn't really looked. You know how it is when things conspire against amphibian-related outings. Somehow 'find large newt' kept slipping down my lengthy 'to do' list. While I regularly saw endearing little palmate newts in the village pond, with distinctive dark webbed back feet and thin filaments sticking out of the ends of their tails like stripped fuse wires, the distribution of great crested newts in the area where I live was largely unknown. A bit of travelling was required. This was the time of year when adult newts that had been lying low over the winter in burrows and crevices finally returned to their favoured ponds to breed. I felt certain that joining a survey training day organised by the charity Froglife, which included visiting a nature reserve with arguably the world's biggest population, would guarantee an easy sighting. Only I hadn't reckoned on the cold weather. Course leader Paul Furnborough warned me that low temperatures meant they could be either inactive or in hiding, and said some 'expectation management' was required.

Nevertheless I decided to take a chance. Leaving home at 11pm after a late work shift, I drove through the night in order to reach Peterborough in time, passing gritting lorries and piles of old snow along the way. I managed to get a little sleep in a lay-by, shivering on the reclined driver's seat under a duvet, and as the sun came up I saw how the landscape had changed from rolling hills and narrow lanes to vast fields and open vistas. Finally I pulled up outside the meeting place on the outskirts of town at 9am, stiff-legged and bleary eyed. As instructed, I located a woman wearing a 'frog-shaped woollen hat', who showed me inside, where more than a dozen conservationists and volunteers, having given up their Saturday to learn about great crested newts, were gathered.

Much interest in the species stems from its highly protected status. Strange as it might sound, you could be arrested for simply picking up a great crested newt, not that most people would want to. And laws that prevent them from harm have on occasion set these pimply pond-dwellers at odds with developers, delaying building projects while sites are surveyed and measures put in place to safeguard or relocate colonies.

Of course, it is only because we have destroyed or fragmented so much of their habitat in the first place that they have earned such protection. It is estimated that we have lost a third of our ponds over the last fifty years. Many have been drained or filled in, particularly on farmland, or have been simply neglected and dried out. Those that remain may be affected by pollution, fertiliser run-off from neighbouring fields or the introduction of fish, and as the newts don't much like crossing roads and seldom travel far from their breeding ponds – one kilometre would be a major adventure – they can become confined to small areas where they are increasingly vulnerable. In addition, a disease known as chytrid fungus, which is wreaking havoc with amphibians around the world, has arrived in Britain and may add to their problems. So it was heartening to be in the company of people keen to improve their outlook.

'They're wonderful to see, and something to be proud of given that Britain is a stronghold for them,' Paul said. 'On land they look a bit beached, but in water they are in their element, really quite majestic, especially when you witness the males displaying.'

Such a spectacle can't be seen in any old stretch of water. For a start, the newts tend to be found in southern, central and eastern areas of Britain, and in small lakes or fairly large ponds with sunny, weed-free areas where the crested males can show off in the spring, waggling their flattened tails and fanning pheromones towards passing females in the hope of enticing them to mate. They also need a decent supply of invertebrates to eat and plenty of water plants on which the sticky eggs are laid, individually wrapped in the folded leaves like Greek dolmades. Those young that hatch take a few months to develop fully into smaller versions of their parents before leaving the pond, and they need plenty of rough cover on land because, perhaps surprisingly, great crested newts spend most of their lives out of water, taking several years to reach maturity. This is also where they are likely to be confused with lizards, although you could say there is a hard and fast rule for telling them apart: scaly lizards are just that – hard to touch and fast when active – while velvety skinned newts are soft and slow.

I was keen to get searching. However, I had to be patient because 'cresties' are most active after dark. There was a day-long course on newt ecology and surveying techniques to get through before we finally set off at dusk in convoy for the prime site.

An industrial wasteland lying between a dual carriageway and a new housing estate to the south of Peterborough doesn't sound like the most promising setting for wildlife watching, despite developers having named the final stretch of asphalt leading there 'Nature's Way'. But, incredibly, nature has found a way in this unlikeliest of places. For over fifty years the area we had come to was a playground for diggers. Thick clay deposits were gouged from the ground for use in the mass manufacture of bricks, leaving a lunar landscape of spoil heaps and trenches. Yet since excavations ceased in the 1990s, the brownfield site has turned a healthy shade of green, and water that collected in the deep furrows is today a haven for aquatic organisms, including rare plants, insects and, lurking black as lawyers' robes in the old pits, enough great crested newts to bring an urban planner out in a cold sweat.

The land is now managed as Hampton Nature Reserve by Froglife on behalf of the owners O&H Hampton, but is not open to the public. The three hundred steep-sided pools that make the location so good for wildlife also present a hazard for unwary visitors and, once we had parked, Paul gave us a health and safety briefing on the former quarry's potential pitfalls, in the literal sense of the word.

'Oh, and don't lick the newts,' he advised.

Skin secretions of great crested newts contain a weak toxin, making them unpalatable to some predators – and humans as well. We know this thanks to the brave, if bizarre, experiments of the eminent Victorian naturalist Eleanor Ormerod, who, 'for the sake of ascertaining the sensations', gave a newt a gentle nibble. At first her mouth became numb, then began foaming, and then underwent 'violent spasmodic action, approaching convulsions', after which she suffered headaches for several hours and shivering fits. Her cat was also roped into the taste test and didn't appear to fare much better.

After everyone had disinfected their wellies to prevent any possible transmission of diseases onto the site, it was time to put training theory into practice. We had come equipped for the search, with fine mesh nets, traps made from spliced sections of plastic water bottles fixed to lengths of cane, and powerful torches. But as the sun sank and the temperature plummeted I wasn't optimistic that any cold-blooded creatures would have the energy to stir. It was freezing. I couldn't see any eggs, no one caught anything in the nets after half an hour of trying, and the only thing moving on the bottom of the pools seemed to be caddis-fly larvae.

While the others practised setting bottle traps, I joined conservation worker Becca Neal – the one in the froggy hat – to check a few good spots on land. On our way she pointed out a cluster of hard objects like thumbnails in the clay: fossilised shellfish from the Jurassic period, when this area lay under the sea. Curled up beneath a piece of wood were a couple of common lizards. We could also hear water birds settling down for the night. All the great crested newts, it seemed, were hiding.

Having examined likely areas, we checked a forty-foot length of abandoned conveyor belt by the track, just in case anything was sheltering under it. Lifting the thick material bit by bit, we scanned the pale flattened grass with no luck, until something caught our eye at one end, tucked up in a twist of vegetation. It would have been so easy to miss. Becca wet her hands in a nearby pool and lifted it up to show me. There it was, stretched across her palm: a striking female great crested newt. She was so much larger than any newt I had ever seen before. A real heavyweight, with bright eyes, dark skin flecked with tiny white dots along the flanks, a spotted orange belly and red rings of colour on her fingers and toes.

For an amphibian she had real presence, and in the way she slowly looked around and took things in, an unexpected charm. Covered by the tutor's licence, I was even able briefly to hold her before she was safely returned to the same spot.

The light had faded by the time we joined the group again. Torches were switched on and strong beams played through one of the pools, from the soft clay edges to the deeper centre thick with weed. There is something absorbing about peering into small clear bodies of water, like inverted glass paperweights containing life. And now darkness had fallen, animals had moved out from cover: water beetles, aquatic snails and a smooth newt that was caught in a net – a handsome male with spotted sides. There were also glimpses of something larger: the dark tail of a great crested newt disappearing in a puff of sediment. They were proving elusive, and while I had enjoyed a decent sighting earlier on, others in the group hadn't been so lucky.

Eventually it was time to leave. Then, walking back to the cars, someone called out from beside a pond up ahead and we rushed over to add our torch beams to his. Against the bottom in full view, with spiky crest and big blade of tail adding to its size, was a superb male great crested newt – a dragon in his lair. It was a splendid vision to conclude a fascinating day. And strange to think that while house building has contributed to the decline of this amphibian, I had brick pits to thank for two memorable encounters.

A slender plume of smoke rose from the headland, snaking up above the treetops into the pale blue sky. It could be seen from the hills to the west, and from across the bay. Somewhere within the forest that cloaked this rolling coastal landscape, a fire had been started deliberately.

The source was hidden from view, but the sound of voices, audible above the crackling of burning timber and the hammering of stone on wood, led the way along tracks that ran through the woodland. The paths converged on a clearing. On one side stood a simple mud and thatch dwelling, the earth around it flattened by the soles of naked feet. On the other, a small group of perhaps a dozen men and women, absorbed in the task at hand. Sweating in the heat, the men, stripped to the waist, skin peppered with ash, were hacking at stands of birch, elm, hazel and oak using axe blades of polished flint. Beside them, the women, wrapped in animal furs, stoked fires at the bases of the larger trees.

It was several thousand years BC, and if you were looking for scarce mammals in Britain at that time you would have counted yourself extremely lucky to have chanced upon this Neolithic gathering. Wolves, wild boar, beavers and bears were a cinch compared with finding humans, whose scattered population was thin on the ground.

But that was about to change, and this group were among those leading the way. After millennia as hunter-gatherers, people here were finally ditching the nomadic lifestyle and settling down. The land bridge that joined our country with the Continent after the last ice age was long gone, but seafaring traders from Europe kept us connected to the outside world. They brought domestic livestock and grow-your-own expertise from further afield that was to transform the way we lived. Wandering around eating berries and barbecuing wild auroch was *so* last year. Farming was the way forward and, armed with simple tools, the inhabitants of Britain waged war on the vast wildwood that stretched the length of the country, clearing the land bit by bit to plant crops, graze animals and put down roots.

On the peninsula now known as the Isle of Purbeck, this group of Stone Age trailblazers had breached the forest fortress. Slash-and-burn tactics enabled them to grow wheat and barley. However, the sandy soil in this part of Dorset rapidly gave up its goodness and eventually the area would be turned over to rough grazing. Heather and gorse then colonised the exhausted earth, providing a source of fuel, while foraging cattle and sheep nipped tree saplings in the bud, helping prevent the open expanse from returning to woodland. The heathland habitat that resulted, with its dry turf and scratchy shrubs, became a feature of our countryside where conditions allowed, particularly in the south and east. Created and sustained by the activities of mankind, it also provided a home for a unique assortment of wildlife.

Today much of our lowland heath has been lost to neglect or development, ploughed up and enriched with fertilisers for farmland, or converted to conifer plantations. It is estimated that eighty per cent of heathland commons have disappeared over the last two hundred years, and what remains is now recognised as being of international significance, widely protected for nature and its recreational value. But it takes a lot of looking after. Like most of the land in Britain, it yearns to return to its wildwood past. Preventing that from happening means adopting a tough love approach: cutting, burning, grazing, bashing, slashing . . . whatever it takes to reduce soil fertility and keep scrub, bracken and trees from encroaching. At the RSPB's Arne reserve near the Purbeck Hills on the edge of Poole Harbour, visitors are even invited to join the effort on winter 'pull a pine' days. Hundreds turn out to dig up and take home a free Christmas tree, in the process helping to conserve heathland that has existed here for thousands of years.

It was late April when I visited the reserve – a little too early to start thinking about Christmas, even for department stores. I was in search of a reptile that has become as scarce in the UK as the heathland it depends upon.

Visitor manager Rob Farrington offered to act as my guide. The sun was out, and it was shaping up to be the first mild day of the year. Reptiles are most easily seen when they are warming up – too hot and they're off before you even know they're there – so we got going as soon as I arrived.

The route we took passed through a farm area where a few rectangles of corrugated iron had been laid on the ground in a scruffy corner. The metal sheets absorb heat quickly and attract reptiles wanting to raise their metabolism for the morning ahead, or top up their temperature later in the day. We lifted them gently to find two slow worms sheltering underneath, smooth and shiny as coils of brass pipe. Dark sides and thin black lines running down their golden backs indicated they were females, and Rob estimated that the larger of the two was eight to ten years old, which was a fairly good age for a slow worm in the wild. They were lovely to see, but not the reptiles we were after.

From the farmyard, the path ran along the edge of a field planted with a bee-friendly mix of flowering crops, then cut through oak woodland, where I managed to put a few names to birds. The easy ones, anyway. The appeal of nature watching can be as much about the satisfaction of identification as the joy of observation. Learning to recognise what's around you adds immeasurably to the pleasure of being outdoors, and fortunately birdwatching, unlike hang gliding, is one of those hobbies where it doesn't really matter when you get it wrong. Which is just as well. While I would love to culture an impression of cool expertise, enthusiasm always gets the better of me and I can't help blurting out when I think I know what things are, even in new places where everything appears unfamiliar.

'Look, there's a couple of goldfinches in the hedge.'

'Linnets,' Rob corrected me.

'Oh yes, linnets. That's what I meant.'

We crossed an open area that was dotted with occasional circular pools and depressions. Bomb craters. Puzzling, given that Arne doesn't seem like the kind of place worth shelling. Its fields, eleventh-century church, school house, farm and few dozen cottages were hardly top of the list of German targets during the Second World War. Instead they had their sights set on an explosives factory at Holton Heath a couple of miles to the north west. But they missed for a reason. Arne's position, situated on a low-lying peninsula that nudges through mudflats into the natural harbour between Wareham and Brownsea Island, lay directly under the bombers' flight path, making it the ideal site for a decoy target. Fake blazes using a network of piped paraffin and timed explosions of tar barrels fooled German air crews into thinking this was Holton Heath taking direct strikes. The result was that they dropped their payloads too early, and on one night in June 1942 more than three hundred bombs fell on Arne, devastating the area but saving the factory. Today the craters provide small sanctuaries for wildlife, such as grass snakes, uncommon spiders and carnivorous sundew plants.

Before long, Rob and I reached one of the reserve's large swathes of heathland – acres of dry heather that would become a humming carpet of purple in the summer. This early in the year it looked brittle and dead, but closer inspection revealed signs of life. Sprigs of green were emerging from the woody stems that brushed our calves, and we spotted wood ants and metallic tiger beetles patrolling the sandy tracks.

Rob stopped me to point out a long-tailed bird darting low between two clumps of gorse. It was a Dartford warbler – the reason the RSPB took over this site in the first place. Back in the 1960s, numbers of these little grey and claret birds had slumped to less than a dozen pairs in Britain, prompting the charity to buy up Arne's prime heathland in a bid to save them from national extinction. They have since bounced back from the brink, helped by the protection of this vulnerable habitat on which they depend.

'Of course you can't just manage the reserve for an individual bird, you have to maintain it for a broader range of wildlife,' Rob said. 'And by cutting and burning in rotation we create a mosaic of heather and gorse of differing ages, which supports insects and reptiles and so on.'

If managing for wildlife sounds like meddling, then it is worth considering just how much of the British countryside we know and love is man-made: tamed and fashioned down the centuries for our purposes. Once buried beneath forests, these islands were opened up to the skies by our ancestors, and the variety of habitats we now enjoy – from wildflower meadows and patchwork fields to sheep-clipped hillsides and grouse moors – is the legacy of generations of human endeavour. Our past and present is sewn into the fabric of the land. To step back and let nature take its course could be an interesting experiment, but it would risk the loss of much of the diversity that adds such character to our country – Dartford warblers included.

We had reached an area of the reserve called Grip Heath, where squares of corrugated iron, slightly smaller than those concealing the slow worms earlier, were hidden scattered in the heather. These metal magnets for cold-blooded creatures enabled Rob and his colleagues to get some idea of reptile numbers, and because of the possibility of interference from visitors, this section was closed to the public. Rob got to work, in turn lifting each iron sheet by its edge, like a trap door to a hidden world of secretive species.

There were plenty of slow worms about, of varying sizes. It isn't easy to see where legless body meets tail on something that looks all tail, but it is worth knowing if ever you handle a slow worm. Grab one by the wrong end and you could find it coming off in your hand: they shed their tails to escape, like lizards (and that's because they are lizards, even if commonly mistaken for snakes). Another distinguishing feature is that slow worms have eyelids, though I'm not sure exactly how close we're expected to get to snakes to check whether or not they're blinking.

The corrugated iron squares turned up a few surprises: a fat toad looking a bit out of place and a very light-coloured common shrew. And there were plenty more slow worms. But no sign of the shy southerly species I had come to see.

'Don't worry, one should turn up eventually,' Rob said.

I admired his confidence – and the fact that he seemed to know where all the corrugated iron sections were, despite it being his first survey outing of the year. Still, I didn't want to take the risk that he might miss a single square: 'There's one,' I pointed out helpfully. 'And another deeper in the heather over there . . .'

'Sure, I'll get to those in a minute.'

By now the sun had climbed high into a clear sky and it was surprisingly hot. Parched shrubs, soft sandy trails, common lizards, the coconut smell of yellow gorse flowers and glimpses of dazzling sea between pine trees gave the place a Mediterranean feel. There were even a few spoonbills in one of the distant bays. Hard to believe this was less than three hours' drive from my soggy stretch of Dartmoor.

Rob donned a hat, I swigged from my water bottle and we pressed on into another section of Grip Heath to check a further dozen or so squares.

Slow worm. Next had nothing. Iron sheet blown out of place – nothing. Common shrew. Nothing.

And then Rob lifted the edge of a square wedged in deep heather. Tucked up at one end among tangles of matted roots and lichen was something far larger than a slow worm.

'Got one,' he said.

Two magic words. I could have punched the air and yelled 'Yes!', but tried to keep calm. Then again, why bother? 'Yes!' I exclaimed, unable to help myself as I leaned forward for a closer glimpse of the precious find. And what a find it was: Britain's rarest and most elusive reptile – the smooth snake. And a big one.

Some people are lucky when it comes to snakes – lucky in that they see them regularly. I'm not. While I have come across a few overseas, they have pretty much given me the slip in Britain, even though I've visited all the right kinds of places. Perhaps I have spent too much time looking up for birds, rather than down at the ground, but whatever the reason my sightings had amounted to a small adder near the coast in Dorset when I was sixteen and the tail end of a grass snake disappearing into dense undergrowth in Sussex a year later. Given that dismal tally, the odds seemed stacked against me coming across the third of our native trio: the extremely scarce smooth snake. So it was a mighty relief that my decades-long drought had ended – and with a species that even the most charmed snake spotter would feel fortunate to encounter.

Being one of a select few people in Britain licensed to handle smooth snakes in the wild, Rob deftly picked it up to allow me a closer view. It was a handsome male, and from the pattern of markings on the back of its neck he recognised it as one of just eighteen individuals that had been found on Grip Heath in the past.

'He's certainly grown,' he said. 'This is about as big as they come.'

It looked to be more than two feet long. Perhaps not huge in snake terms – smaller than a typical grass snake, for example, and slimmer than an adder – but a good size for the species, and perfectly capable of tackling lizards and small mammal prey.

Its buff-grey body was subtly marked with flecks of ginger on the sides and rusty squares that ran in twin tracks down its back. The top of its head was capped with a chocolate mark in the shape of a heart and a dark horizontal stripe cut through each eye – wide eyes that gave it a slightly baffled expression.

'Grass snakes may grow large, but they're wimps when you handle them, hissing, pretending to be dead, ejecting a foul-smelling liquid and so on,' said Rob. 'Adders give you a look that says: "don't even try". With smooth snakes, some are placid and happy to steal the warmth from your hands, while others are more feisty and show a lot of character – like this one.'

The snake was weaving between his palms and flattening its head in an attempt to look as if its jaws were bulging with large venom glands.

'Would you feel confident handling it?' Rob asked.

I nodded, and almost looked convincing.

He gave me a few instructions and then passed the snake over. It was dry and light, a slender length of muscle that slipped from hand to hand, drawing loops through my fingers as smooth as silk – a tangible reminder of how it got its name. While our other two species of snake have slightly ridged scales that provide added traction when moving on the ground, the smooth snake has unkeeled scales, enabling it to glide through the mature heather in which it basks and hunts.

I felt it was relaxing in my grasp. Perhaps I was a natural – a snake whisperer in the making. And then, as if to remind me not to get too carried away, bam! It bit me. Right between the thumb and forefinger.

The smooth snake is not venomous. It uses its tough teeth to grab prey before wrapping itself around its victim like a constrictor to prevent escape. But those teeth were certainly sharp and had successfully pierced my skin. It didn't appear to want to let go.

Ouch.

I realised I had only one option . . .

So that's the story of how, with a quick thwack, I came to own a wonderful smooth snake belt.

Er, no. There was no way I was going to hurt this impressive individual, no matter how much he was hurting me. Even flinching, pulling the jaws free or dropping the snake might have harmed it, and I wasn't going to take that risk. The only choice was to grin and bear the pain, and after this big old-timer had made its point, it unclasped its jaws.

I held the snake for a little while longer before Rob released it back underneath the corrugated iron. Having no luck with any of the other squares, we headed back to the car park, my head spinning with the thrill of the encounter. To see a snake at all was exciting enough, but to have got to grips with our rarest – or rather for it to have got to grips with me – was incredible. And funnily enough it felt like quite an honour to have been bitten. I only wished it had left a visible scar – one that people might ask about.

'Oh that? It's nothing really. Just a snake bite, that's all . . .'

# Four

Working in the production department of a daily city newspaper, filling blank pages against the clock, can feel a bit like packing a moving lorry. You start the morning with a virtually empty space, save for a few adverts stacked up in the corners, and push boxes of text into place, sometimes pulling them out and replacing them, holding the best for last, until, breaking into a sprint as the late-night deadline approaches, you squeeze in the final story, slap the unstoppable juggernaut on the back and off it speeds, whether you like it or not.

There was a time when plenty of production staff accompanied me on that daily race, designing pages, editing content, writing headlines and proofreading. But times have changed. Readers and advertisers have increasingly abandoned print for the Internet, and waves of redundancies have swept through the newspaper industry, carrying away staff and leaving fewer and fewer journalists clinging to their desks. Career-wise I have been in as much danger of extinction as some of the animals I hoped to see.

Staff losses made it increasingly difficult to sneak the odd weekday off for nature watching without being noticed. And faking an illness was out of the question, on account of my poor acting skills. However, working Sunday shifts earned precious time in lieu, and this enabled me to arrange a Thursday away from the office at the end of April.

'Got anything planned?' a colleague asked.

'I'm joining a wildlife survey group in Dorset,' I replied.

'Oh, that sounds interesting.' (Neighbouring staff had become proficient actors.)

'Yes, looking for bats.'

'Bats? Hmm . . . I can't really see the appeal.'

I'll be honest, I was struggling with the idea of bats myself. I knew that I was supposed to embrace them like everything else but, with their screwed-up faces, leathery wings, nocturnal habits and sinister reputations, I needed some convincing. It was nice to see one fluttering past every now and then, but that had always been enough for me. If I had to see a rare bat, I would make it a quick trip and get back to more loveable, fathomable creatures, I decided.

And what a freaky bunch of British bat species I had to choose from! I guess nothing needs to look its best in the dark. Grimacing heads moulded for night navigation stared out of the pages of my mammal guidebook, wings like broken umbrellas, thin back claws trailing, furry bodies looking anything but cuddly. Some had outsized ears or bald muzzles, others squashed noses or jaws bristling with little teeth. Not one looked like it wanted its picture taken.

Despite their slightly alarming appearances, we have nothing to fear from the eighteen types of bat found in Britain. For a start, all of them are insect eaters, with not a bloodsucker among them, and they are small – our common pipistrelle weighs about the same as a twopence piece and could fit inside a matchbox, even though its open wings make it look far bigger. And regardless of the cautionary tales, no bats are such incompetent fliers as to make a habit of getting tangled in people's hair. If that were true, then cases would surely have soared in the 1980s, when the vogue for backcombing and big hairstyles presented a real obstacle for any low-flying species – and I can't recall people queuing up outside hospitals and hairdressers at the time for emergency bat removal.

If bats occasionally do swoop close to our heads it is simply because they are chasing insects around us, and they need plenty of those to keep them going – a couple of thousand a night in some cases. There may be a smorgasbord of moths, bugs and mosquitoes on the wing during the summer, but taking advantage of all that airborne food comes at a price: flying burns up a huge amount of energy. So much so that during poor weather, resting bats save what they can by allowing their temperature and metabolic activity to fall, slipping into a torpor. And throughout the long cold months of winter, when insect life has dwindled, they take this survival strategy to the extreme and hibernate, like dormice and hedgehogs.

One of the reasons our bats can look so odd is that their strangely shaped noses, mouths and ears have evolved to act as transmitters and receivers. No bats are blind, but all of them rely on echolocation to find their way and to pinpoint prey after dark, emitting high-pitched noises and interpreting the strength and direction of the reflected sound to map their surroundings. This incredible ability has given us such peculiar specimens as the greater horseshoe bat, with a call-emitting nose that appears to be inside out; the pug-faced barbastelle bat, which looks as if it has flown into a window; and the brown long-eared bat, able curl up its huge ears like ram's horns when not in use.

The rare species that I intended to see was among the least photogenic of our residents: Bechstein's bat. Named after a German naturalist and not the famous piano maker, this woodland dweller, with its virtually hairless pink face and elongated ears, was once abundant in Britain. However, the loss of ancient deciduous forest has resulted in it becoming one of our rarest mammals. Today, Bechstein's bats are largely confined to southern England, where the population numbers perhaps 1,500. It's hard to know exactly how many there are, because so few hibernation sites and summer roosts have been found. And its whispering ways also mean that individuals are notoriously difficult to pick up in flight using bat detectors. These hand-held devices convert bats' ultrasonic peeps, warbles and squeaks into sounds within our range of hearing, and enable users to match the frequency with the species. Except that Bechstein's bats don't play along. They hunt by stealth, listening out for insects in the canopy and keeping calls to a minimum.

My best hope of seeing one of these quiet, scarce and mysterious bats was to join a group visit to an extensively monitored stronghold at Brackett's Coppice Reserve in Dorset. In this sloping area of woodland, expert Colin Morris, from conservation charity the Vincent Wildlife Trust, has successfully persuaded Bechstein's bats to take up residence in a number of special boxes. And they now come back year after year to raise their young.

However, the long cold spring spelled potential disaster. After weeks tucked away in hibernation refuges, our bats, Bechstein's included, had woken with depleted fat reserves, only to find little in the way of breakfast on offer. Many were known to have died. And those females that had survived needed desperately to put on weight before assembling in summer maternity colonies and giving birth. It was anyone's guess whether they had returned to Brackett's Coppice.

I left home at 7am, which gave me three hours to reach the reserve, factoring in plenty of time for getting lost. The truth is I have a lousy sense of direction, an appalling memory for place names and a short attention span when it comes to being told the way to go. Journeys that start out following neat straight lines frequently end in a tangle of wrong turns, circular detours and desperate searches in second gear with the window wound down for anyone who looks local. And it was just as well I had set out early. A dense morning mist obscured landmarks and road signs, as if a spade had been wedged under Dorset and the county levered up into the clouds, and it took a short tour of B-roads before I found the entrance to the reserve, where the group had already assembled.

So what kind of people are bat enthusiasts? The answer seems to be a perfectly normal and friendly bunch, judging by those I was joining. There wasn't a goth among them.

Colin got proceedings off to a humorous start with a health and safety briefing that listed all the various ways we could expect to die upon entering the wood, including attack by wasps, wild boar, drowning in the stream and either falling from trees or having trees fall on us. Then, hard hats at the ready, we got going, because there were dozens of bat boxes to check. We took a muddy path that led between the stands of oak and ash. The wood anemone was in flower, as well as the charmingly named Goldilocks buttercup, and I heard a great spotted woodpecker drumming. We also passed patches where the leaf litter had been pushed aside and a scattering of brittle hairs indicated roe deer had slept there.

It required a ladder to reach the bat boxes, which were wired two or three to a trunk, although heights were not a worry for Colin. Originally he worked as a roofer, renovating old buildings, which is how he first developed an interest in bats, enthusiastically reporting those he came across to environmental officials before eventually becoming a conservation worker himself.

'The thing that appeals to me the most is the opportunity to handle and conserve species that are alien to ninety-nine per cent of the population, and in some cases extremely rare,' he said. 'Bechstein's bat is certainly that. It isn't particularly attractive, in fact it's a bit of a gremlin to look at, but it is an intriguing and enigmatic species.'

The magic moment for Colin came back in the 1990s. Up until then our Bechstein's bats had refused to take up residence in standard wooden bat boxes, sticking with natural holes and crevices in old trees, where they were virtually impossible to find. However, researchers in Germany had successfully used domed boxes constructed from a durable mix of concrete and sawdust, and Colin arranged for a number to be imported and tested out at various locations.

Years passed before any Bechstein's bats showed the slightest interest. Half a dozen males found in a few boxes on the Wiltshire border offered a ray of hope, but not quite grounds to crack open the Champagne. Then, in June 1997, during routine checks, Colin opened box number '2' in Brackett's Coppice and could hardly believe his eyes. 'I had to restrain my excitement because I was at the top of a ladder,' he said. 'I brought one down to show a colleague and said calmly that I thought I had found something quite interesting. When they realised what I already knew, that it was a female Bechstein's bat, I added: "Oh, and there are sixty of them in there."'

This was Britain's first ever colony of breeding Bechstein's bats to be discovered in a bat box. It enabled Colin to keep detailed records of this rare species over the years that followed, fitting offspring with metal identification clasps, similar to those used on birds. Among them, his favourite four 'old ladies', as he likes to call them, have been putting in an annual appearance, with one female raising at least nine young at the customary rate of one per year.

'She must sigh every time I open the hatch and think: here he is again!' he said.

Climbing the ladder, Colin checked the first bat box of the day, prising open the front section and peering inside.

'Bats present,' he said. Not Bechstein's bats, but brown long-eared bats, one of six other species found in the wood. He brought one down for a look.

I was ready to force a smile and make a few appropriately positive comments, as one might about a particularly ugly baby, but it was actually quite sweet looking – half asleep and trying its best to look fierce by baring its teeth. The tiny canines would struggle to puncture skin, though people who handle bats wear gloves just in case. They also keep up to date with their rabies jabs, because bats are able to fly across the Channel and rare cases of individuals carrying a strain of rabies virus have been recorded.

As the bat warmed up, its ears unfurled, which prompted a collective 'ahhh', before it was returned, still fairly torpid, to join seventeen others huddled together between the wooden slats.

Most of the boxes we checked next were empty, including number '2', although deposits of dry, crumbly droppings indicated others had been visited. A couple of boxes looked as if they had been taken over by nesting birds. Eventually we found one occupied, this time by another small species: the Natterer's bat. Eight were visible, their petal-shaped ears showing and their wings folded like crutches on either side of their bodies – only our lesser and greater horseshoe bats hang upside down with wings wrapped around them Dracula-style.

A box deeper in the wood turned up another species – and one that required thicker gloves. 'Can you hear them?' Colin asked the group, as he climbed up the ladder. The squeaking was clearly audible from the ground four metres below. 'Noctules.'

Large, fast and high-flying, noctule bats are a serious handful, and there were about fifty in the box. But Colin managed to extract one and gave us all a closer look. It was a lot bigger and tougher-looking than the previous two bats, with a dark, brutish face and a body covered with uniform golden brown fur, as if it was wearing a jumper.

Those with bat licences helped Colin check the boxes, and they turned up a few more of the species already seen. All I could do was stand underneath the ladder and hope that the last half-dozen boxes would come up trumps. But we had no joy with the first couple, nor the next two, an empty box followed and then came the last one, and . . . nothing.

It was disappointing, but sometimes you have to earn the right to see rare species, and setting out with an ambivalent attitude towards bats had loaded luck against me. Why should I simply swing by and see a Bechstein's bat when so many genuine bat lovers went a lifetime without encountering one?

Yet something had changed in me during the day. I had begun to warm to those dozy palm-sized balls of fur we had found, with their toothpick-thin wing bones and hangover expressions, and to appreciate the differences that gave the species character. In all, eighty-three bat boxes had turned up a total of 130 noctule, Natterer's and brown long-eared bats – a remarkable tally for such a small wood, with plenty more still due to settle in for the summer. The visit had opened my eyes to the diverse night life that we miss when our eyes are shut. I had a lot to learn about bats, but for the first time I actually wanted to know more.

When it comes to naming animals, alliteration is virtually compulsory. So an anonymous male Bechstein's bat that turned up at a rescue centre just had to be called Brian. Brian the bat. That's the law.

The life of Brian had almost been a short one. At only a few weeks old he had been seriously injured when a falling tree branch smashed into a bat box at Brackett's Coppice in 2012. His shoulder was in such a poor state that it was certain he would never fly well enough to survive in the wild, and he was taken into care.

Brian, believed to be the only Bechstein's bat living in captivity in Britain, now resides with bat warden Lizzie Platt at her hilltop house in Ivybridge near Plymouth, sharing his enclosure with his best pal – a pipistrelle bat called Ernest. (Ernest? Surely it should be Peter the Pipistrelle.)

'I was terrified of bats until about ten years ago,' Lizzie told me. 'It was when I visited a friend of my mother's who looked after bats that I saw how small and fascinating they were and realised it was ridiculous to be so afraid.'

An eager convert, Lizzie now looks after rescued sick and injured bats, on top of her warden duties checking roosts in properties. When I called, she had twelve in her care, a number of which were housed in a wire-mesh pen in a spare room. Opening it, she gently lifted out the common pipistrelle Ernest, whose dark ginger fur-covered body hardly spanned three fingers. He was absolutely tiny, his eyes were closed and he was quivering like an idling car as he tried to warm up and rouse himself from his torpid state – a process that could take twenty minutes.

'He's so sweet, don't you think?' Lizzie said.

I had to agree. Sweet – and strange. Bats certainly appeared complicated little creatures, able to raise and lower their heartbeats, to see with their ears. Not at all as straightforward as mice, with which they are often compared (in fact bats are more closely related to us humans than to rodents). There is something both prehistoric and futuristic about them, with naked wings out of the Jurassic era and space-age echolocation abilities. How on earth did this living copper battery, charging up in Lizzie's hand, experience the world around it? What did we look like in sound?

Half awake, Ernest appeared as befuddled as one would expect after sleeping upside down. Lizzie gave him a stroke and he seemed happy to be held.

'Bats are social animals and not only groom each other but also forge friendships, even with different species,' she said. 'If I take this pipistrelle out on an education visit, when I put him back, Brian will come to greet him and they'll play together.'

My intention was to see rare species in the wild, and I would persevere with my search for such a Bechstein's bat. But ever since hearing about Brian, I had wanted to pay him a visit.

Placing Ernest back in the pen, Lizzie found Brian and lifted him up to show me. 'Perhaps he's not the most beautiful of bats, but I still find him very appealing. He has a really playful personality,' she said.

Actually, he wasn't that bad looking. Perhaps a little more fur on the face wouldn't have gone amiss, or slightly more proportionate ears, or smaller nostrils or . . . no, he was fine just as he was. Bigger than the pipistrelle, yet still small, with dense fur that was soft to touch and lighter on the underside. One wing joint was noticeably swollen, but apart from that he appeared in good health – and when it comes to diagnosing broken flight bones, bats must make life easy for vets, because their wings are virtually X-ray images as it is.

'It feels like a real responsibility and a privilege to be looking after him,' Lizzie said. She was obviously doing a good job. Plenty of affection, stimulation and mealworms were keeping him going, and he was gradually learning to fly – though generally in a downwards direction.

We spent a bit of time watching Brian, before she returned him to the enclosure. I was delighted to have met such a rarity, and even more keen to find one in its natural setting.

Only, I had another animal to see first, now the weather had turned milder – a solar-powered streak of green energy.

May had arrived, and after the cold start to the year that held everything back in March and April, it felt as if the brake cable had finally been cut as the days hurtled headlong into spring. Skylarks on the moor rose from the grass in unbroken song, their trembling syrinxes sieving the air into thousands of semiquavers. Cuckoos were back in our valley, always sounding like something from a bygone era, an echo from the 1700s. And butterflies were on the wing as the days warmed, welcome snippets of colour.

The weather was no longer just a conversation starter; it had come to mean the difference between success and failure on my trips to see rare animals, and my mood increasingly mirrored the forecasts. So, with the disc of yellow on my computer screen matching the one rising outside the window, I set out with a sunny disposition on another heathland visit, this time to north-east Dorset in search of the sand lizard, Britain's second scarcest native reptile after the smooth snake – which, incidentally, eats sand lizards. (Then again, smooth snakes also eat baby smooth snakes. And sand lizards eat baby sand lizards. All of which means the pair of them seem determined to make themselves as rare as possible.)

Town Common near Christchurch, one of dozens of reserves managed by the Amphibian and Reptile Conservation Trust, spans enough quality heathland habitat for populations of both species to thrive. My visit provided the perfect excuse for the Trust's chief executive, Tony Gent, to leave his office and get out spotting the wildlife he works to save. He came equipped with binoculars for scanning foliage and a couple of long-handled paint rollers with the sponge sleeves removed, which, he explained, were ideal for hooking under corrugated iron squares to safely check underneath them, as there were plenty of adders about.

'How would you rate our chances?' I asked, as we set off down an old railway line trail that cut through woodland to the heath.

'Never expect to see a reptile,' he said. 'That's the general rule.'

It was shaping up to be a lovely day, with a clear sky and not a breath of wind. The lizards, Tony told me, tend to bask in patches of sunlight close to cover into which they can disappear in a flash, and spring is the time to see the robust breeding males at their very best, their flanks gleaming vivid green.

'They are cracking looking, real jewels of the heathland,' he said. 'They can look so exotic it seems as if they don't belong here – but they definitely do.'

As the name suggests, sand lizards live in sandy places, such as dry lowland heaths and grassy dunes where they can tunnel into the ground to hibernate or bury their eggs. Once more widespread, their range contracted significantly during the twentieth century, as suitable habitat was lost to housing, agriculture, forestry and leisure development, leaving just a few scattered dots on the distribution map in Hampshire, Merseyside, Surrey and Dorset, including the area we were visiting. However, reintroduction projects led by the Amphibian and Reptile Conservation Trust have seen captive-bred lizards returning to former haunts, helping to secure their future.

They may also be just the kind of sun-loving species that global warming could benefit. But before four-by-four owners start sporting 'I'm doing it for the lizards' stickers in their rear windows, it is worth noting that climate change may not be all good news for reptiles. Mild winters can keep species active, burning up precious reserves when there is little food about and they should be lying low. And even when times are good, rare animals that rely on carefully managed habitat could be left stranded in protected areas, unable to move with the shifting conditions – one of the reasons why conservation bodies are now trying to break out beyond reserve boundaries and improve landscapes for wildlife on a far grander scale.

We followed tracks through the dry heather, checking the south-facing fringes of foliage. Every now and then I heard rustling and glimpsed something disappearing from view, a brown tail slipping into the cover like a thin tongue retracting. Lizard spotting could be infuriating. But Tony was looking further ahead for reptiles we hadn't yet disturbed, and soon pointed out a common lizard basking beside a clump of gorse. He also checked for snakes under a few corrugated iron squares that had been hidden amid the scrub, turning up a slow worm or two, while I spooked a nightjar, a secretive nocturnal bird that took off from beneath my feet.

I was growing to love heathland, with its spongy turf, spiky foliage and the wildlife secrets it kept hidden close to its chest, concealed among the low shrubs and latticework of roots that threatened to trip me up. Pushing between the heather with head bowed, scrutinising the ground, I felt as if I was wading deep into this place, rather than merely walking through it. I trod heavily though, conscious that there were adders lurking in the shadows, and it wasn't long before we spotted one: a light-coloured adult male with a telltale dark zigzag down his back and red eyes that watched us pass by. And soon after, a smooth snake crossed our path – an incredible piece of good fortune. Tony, a zoologist with a PhD in smooth snake behaviour, said he had seldom chanced across one out in the open.

Strange how once you see a species for the first time, you start coming across it again and again. It is more than simply a case of getting your eye in; instead it can feel as though you have earned a special permit entitling you to future sightings. I had claimed my smooth snake pass at

Arne, and now they were throwing themselves at my feet.

As for sand lizards, they were proving a bit trickier: a sliver of tail, a ripple of back, a fleeting hint of green . . . enough segments to build a whole one, but not a proper sighting – until Tony stopped me in my tracks with a raised hand. Beside the path, where the dry sand slumped beneath crusts of coffee-coloured peat, a stationary female was soaking up the sun. Larger than the common lizard, she was covered in a smart, carpet-like pattern of dark brown spots with cream centres, and two light lines ran down her spine where the markings faded, as if removed with a pencil rubber. Then she was gone, darting off between the woody twists of heather.

Tony drew his heel through the sand, marking the spot. Lizards frequently return to the same place to bask, and we would come back later.

But we didn't need to. She was trumped by a show-stopping male – a brief encounter close to the main track. He dashed through a patch of dry grass, a bolt of electric green. Yet when he paused, one could easily have missed him altogether: his body became a length of dried bracken, his hi-vis flanks a smudge of fresh moss.

In all I had spotted five of our six native reptiles in a morning. They read like a shopping list for a witches' brew. In fact, most of our herpetofauna end up in the cauldron in Shakespeare's *Macbeth*, from the 'blind-worm' (slow worm), 'adder's fork', 'toad', 'eye of newt' and 'toe of frog' to the 'fenny snake' – a fenland species likely to be the grass snake, the one reptile I had failed to see. If the coven were after a soupçon of sand lizard for the pot, this heathland was the place to visit. My views may have been fairly brief, but in such good habitat sand lizards can number more than one hundred per acre – although how anyone manages accurately to survey such a hit-and-miss creature I have no idea.

Assessing the status of any animal is seldom straightforward. It begins with a basic question of numbers: counting and mapping and comparing, and this country is fortunate in having a long tradition of recording wildlife sightings and an army of volunteers willing to help gather this valuable information. Once you know what you had, what you've got and why they differ, you can begin to make informed choices about what needs to be done – and what you can realistically hope to achieve. And in the case of the sand lizard, the effort has paid off and things look a lot rosier than they did. You might even come across one in Devon or Cornwall, Kent, West Sussex or North Wales these days. Only why, one might ask, should we care about saving sand lizards? Do we really need them? Would it matter if they had disappeared for good?

From a purely economic point of view, it probably wouldn't. We don't eat them, trade in them or use them for medical purposes, and, unlike crowd-pleasing ospreys and white-tailed eagles, they don't offer much of a return on investment by attracting coach loads of ecotourists. Putting a price on what wildlife does for us might seem a rather self-centred and utilitarian way of regarding life on earth, but in a world where money talks, it is one way of demonstrating to decision-makers the cost of losing species. On a global scale, nature's services have been valued in the tens of trillions of dollars, and we are increasingly being made to pay for mismanaging them. In one area of south-west China, for example, workers have to be employed to hand-pollinate pear trees after the indiscriminate use of pesticides killed off all the local bees in the 1980s.

Sand lizards may not be worth much in a financial sense, but it can be argued that maintaining the greatest possible variety of flora and fauna – reptiles included – is a wise insurance policy. Ecosystems have been shown to function better where there is greater biodiversity, and we can never be entirely certain what impact losing a species will have in the complicated, interconnected scheme of things. Of course plants and animals come and go, and more than ninety-nine per cent of those that have ever existed have died out. However, we are currently losing wildlife at such an unprecedented rate that scientists believe we are bringing about the sixth mass extinction event since life began – the last one being the sudden loss of dinosaurs in the Cretaceous period. Today more than twenty thousand species are considered endangered, including a quarter of known amphibians and one in five mammals and reptiles. Who knows what their decline or departure could mean for us and future generations?

Fortunately sand lizards are not threatened with global extinction, though that doesn't mean we can take them for granted.

'We have a duty as custodians of the countryside to conserve the diversity of wildlife that belongs here, and sand lizards are very much part of our natural heritage,' Tony said.

The value of species extends beyond their usefulness, the physical and material benefits they offer us, the hard currency of biological resources and ecosystem services. Our relationship with plants and animals and habitats is also an emotional, spiritual, aesthetic and cultural one. The connections run deep. The natural world moves us with its myriad solutions to the challenges of life. It resonates within us, even in our modern lives, such that we share our city homes with pot plants and cats, spend a small fortune feeding garden birds, watch TV documentaries about safari parks and penguins and support appeals to save polar bears and rainforests, regardless of whether we may actually get to see or visit them ourselves.

The sand lizard is one small cast member in the earth's great wildlife show, a colourful character that makes our heathlands that little bit more special. Worth safeguarding for its own sake, it is as precious as you want it to be. Priceless. We would be the poorer for its loss. And, for my own part, it felt good just knowing they were there, that I could set out to see them again, and that other people might enjoy coming across them.

# Five

So much for detailed planning.

When it comes to seeing rare animals, it can boil down to being in the right place, at the right time, with the right person. The locations where my next scarce species could be found amounted to a few dozen scattered sites, mostly in southern England, supporting meagre populations. Its short breeding season, delayed by the cold spring, meant I had only a month or so in which to spot one, with wet and windy days a virtual write-off. And given the fact that they are not much bigger than a postage stamp and active only for a few hours each day, I was relying on expert guidance.

After my sunny sand lizard sighting I managed a quick sandwich before heading east, starting out on the two-hour drive to Ramsdean Down in Hampshire, where conservationist Dan Hoare had agreed to meet me. Freshly emerged adults had been seen on the wing a few days previously, the weather was perfect, and if I hurried over I stood every chance of glimpsing Lepidoptera royalty. The place, time and essential help were all coming together neatly. I was off to meet a duke.

The Duke of Burgundy is arguably our rarest and most endangered breeding butterfly. Along with such species as the high brown fritillary and the wood white, it is precariously poised on the precipice of extinction in Britain. It wouldn't be the first to disappear. We have lost five butterflies over the last one-and-a-half centuries – attractive ones at that: the Mazarine blue, large copper, black-veined white, large tortoiseshell and most recently the large blue, which has made a comeback after being successfully reintroduced. As things stand, our species tally runs to fifty-nine, ranging from the familiar whites and impressive 'aristocrats' – such as the purple emperor, red admiral, painted lady, peacock and small tortoiseshell – to the moth-like skippers and petite hairstreaks and blues.

The Duke of Burgundy stands alone among them, a handsome noble of ambiguous lineage. For not only is the origin of this butterfly's stately title unclear, but so are its ancestral connections. Once known rather sweetly as Mr Vernon's small fritillary, after the Cambridge-educated naturalist and collector who first netted one at the end of the 1600s, the species later assumed the grander name Duke of Burgundy fritillary. The reason remains a mystery. While the regal status has endured, its genealogy has come under scrutiny. Despite sharing the chequered orange and black wing markings of other fritillaries, it is no longer considered to be a member of this grouping. Instead, it is believed to be Britain's lone member of the globally diverse metalmark family of butterflies, and its tail-end misnomer has been taxonomically trimmed as a result.

The Duke of Burgundy's identity crisis also extends to its behaviour. It acts as if it were several sizes larger. Gram for gram it is one of our most pugnacious butterflies, a flyweight champion that punches well above its weight as it fights off far bigger species that stray into its territory. A little royal with attitude – you don't mess with the Duke.

The prospect of an audience with such a distinguished species felt like a real privilege, and I could hardly wait. Only there was a potential snag. Dan, a regional officer with Butterfly Conservation, had emailed earlier in the month to warn me that his partner was due to give birth any day. The chances of it being the Tuesday we were scheduled to meet up seemed slim, but I decided to pull over on the A338 and give him a ring, just in case.

He took a long time to answer. 'Hello?'

'Hi Dan, it's Charlie here, Charlie Elder. I'm on my way over. I guess we couldn't ask for better conditions! I should be at the Butser Hill car park in a couple of hours.'

'Charlie? Oh, yes. Um, listen, sorry but I'm afraid I won't be able to make it after all. My partner has just had a baby.'

'Oh dear.'

'What?'

'Nothing, I mean . . . Are you sure?'

'Am I *sure*?'

'No, I . . . yes, congratulations, that's great news. Wonderful!'

'Thanks. I'd better . . . '

'Yes, I'll leave you to it. I'll get in touch another time. Congratulations again. Bye.'

Of all the days. What was she thinking?

I turned the car around and headed home, a tad disappointed.

However, Dan was thoughtful enough to email the details of one of Britain's most dedicated butterfly enthusiasts and a Duke of Burgundy devotee, Neil Hulme, who was happy to meet up a few days later at a South Downs Site of Special Scientific Interest managed by the Murray Downland Trust. The forecast was far from ideal, Neil warned me, though there was a narrow window of opportunity in the late morning, so long as the wind eased and the rain held off. It was a chance I was willing to take, well aware that I could end up driving all that way just to give my windscreen wipers a workout.

Everyone likes butterflies. At least I certainly hope so. Ignoring the damage caused by some caterpillars, they are impeccably behaved and incredibly attractive. They don't sting, bite, buzz, spread disease or get tangled in your hair like all those pesky bats. And they seem to accompany good moods, appearing on sunny days, animating cheerful scenes with colour and movement like dabs of bright paint that have flown off the canvas.

But, just like our summers, they don't last long. Individual butterflies generally live for no more than a week or two. They emerge in waves during a species' flight season, the duration and timing of which varies from butterfly to butterfly, year to year and even region to region. Given adults' short lives and our unpredictable weather, they have little time to waste at this reproductive stage of their life cycle. Males with one thing on their miniscule minds employ differing tactics to secure a mate, from patrolling territories to grabbing a fresh arrival before she has even had time to stretch her wings. Mated females then dedicate their remaining days to laying fertilised eggs on suitable food plants.

The lifespan of a handful of our larger butterflies, however, can extend from one year until the next. While species typically get through the colder months in the form of an egg, a caterpillar in a dormant state or a chrysalis, a few feed up on nectar and overwinter as adults, tucked away in hollow trees and outbuildings with wings folded like dried leaves – occasionally putting in surprise appearances on warm winter days. The metamorphosis from pupa to butterfly may be one of the marvels of nature, but so too is the moment of transformation, when that dark, ragged leaf opens to reveal the striking blue-fringed markings of a small tortoiseshell or the spectacular eyespots of a glorious peacock butterfly.

Is it any wonder that, of all the insects, butterflies have won such a place in our hearts? Delicate stencil shapes of exquisite and ephemeral beauty, sources of wonder and inspiration, symbols of romance and rebirth, they seduce us with a flutter of their wings, flashes of colour and their giddy, teasing flight. Naturalists over the centuries have also fallen for their charms. One of the most intensively studied elements of our native fauna, butterflies have proven the perfect quarry for expert and amateur alike, being relatively easy to catch, identify and display. As a result, collecting became a hugely popular hobby, particularly during the Victorian era, when lepidopterists travelled by train to hotspots where they could play their muslin nets through clouds of dancing species and return home with impressive hauls. Theirs was a time of plenty. And it was not just the diversity of butterflies that kept obsessives entertained, but also the variety within species: unusual markings, or aberrations, provided an added attraction.

Collecting butterflies is now frowned upon, given the declining numbers of many species, and is illegal in some cases. Today's enthusiast carries a camera rather than a net and takes pictures rather than specimens. In many ways, digital photography has revolutionised approaches to watching wildlife. As with so much else, it is not enough simply to experience or remember any more; we are expected to record, to show and share. That's not to say this is necessarily a bad thing, especially if it broadens the appeal of nature. And of all the subjects to snap away at, butterflies can prove immensely satisfying if you manage to creep up on them unawares – which is easier said than done. Mr Vernon, of Duke of Burgundy fame, is recorded as having pursued a restless butterfly for more than nine miles. I was hoping those in West Sussex would prove a little more obliging.

If anyone looked like he could chase down a flighty rarity over distance it was Neil Hulme. A tanned and fit fifty-something oil-industry consultant, brimming with energy, he was far from the pale and gaunt image I had in my mind of a butterfly fanatic. We met beside a flinty field in the charming village of Heyshott and followed a track to the lightly wooded South Downs escarpment beyond. The grey clouds were holding on to their rain, but it was breezy.

'Dukes detest the wind,' Neil told me. However, he still thought our chances were good, and pointed out a lumpy area of grassland on the flank of the ridge that was once quarried for chalk. The hollows offered shelter and were where the butterflies could be found. 'I'm sure we should see one,' he said with a smile.

In any other company a statement like that would have jinxed it. Time to call it quits, spin on my heel and head for home. However, Neil's confidence was reassuring. Not only has he seen all fifty-nine of our breeding butterflies, but he's spotted enough rare vagrant visitors from overseas to push that total to sixty-five. And when it comes to the Duke of Burgundy, he has dedicated more hours than anyone else to saving this downland population. A passionate advocate is just what they need. When Neil first became involved in 2007, the Dukes of Heyshott Escarpment were nearly all gone, clinging on in a patch just a couple of dozen yards across. On a good day you might have seen two. Two! And unfortunately this has become a familiar picture elsewhere.

The Duke of Burgundy was predominantly a species of woodland clearings, but the decline of coppicing and woodland management during the twentieth century meant that flower-filled glades became overgrown and shaded beneath a closing canopy. The butterflies were forced to follow the sun out into open areas, laying their eggs on cowslips instead of closely related primroses, and are now mostly found in patches of scrubby grassland. A good move, you might think. There is plenty of this around, and cowslips are not exactly rare. So what's their problem? Why have scattered population pools kept on evaporating over the decades?

'They are extremely fussy,' Neil said. 'When it comes to their habitat requirements, the devil is in the detail. Nearly right is not good enough.'

Broad-brush approaches to conservation simply haven't worked for this butterfly, which has been squeezed between the under-management of woodlands and the overgrazing of chalk grasslands to support other species. The soft-edged environment they favour needs a gentle touch, a constant resetting of the clock to keep it somewhere between short sward and scrub. And any old cowslip won't do. As pernickety as vegetable buyers for supermarket chains, the females only lay their eggs on thick, green, upstanding leaves in prime condition that tend to grow within humid tangles of longer grass. This knowledge, coupled with the hard work of volunteers, has paid dividends at Heyshott, where a targeted approach has seen Duke of Burgundy numbers turn a corner.

We passed through an area that had been cleared of trees to aid their spread, and finally made it up a steep path to the key grassland site. Standing among the cowslips and orchids, we caught our breath, admiring magnificent views of open countryside.

'The appeal of butterflies is much more than the species alone,' Neil said. 'It's also about the landscapes you visit. Butterflies engender a sense of place – Glanville fritillaries take you to coastal areas awash with sea pinks on the Isle of Wight, and purple emperors mean beautiful broadleaved woodlands in high summer.'

I would never have visited this scenic part of Sussex were it not for the Duke of Burgundy, and I was looking forward to thanking one in person.

Neil checked his watch. It was 11.30am – exactly the time when these late risers tend to appear. Walking ahead of me over the undulating ground, he slowly waved a bamboo cane he was carrying above the foliage to put up any resting butterflies. It was as good as a magic wand. 'There's one!' he called out. Just like that. Amazing.

A small butterfly rose in weak fluttering flight from the grass in front of him and landed nearby. It was a freshly emerged female, a 'Duchess'. Her abdomen was heavy with eggs and her wings, patterned like cracked marmalade glaze, were in pristine condition, spread wide at first in the sunlight, then folding as a cloud passed over to reveal undersides decorated with stained glass-style sections of brown and orange and white. She was extraordinary for such a small butterfly.

Concerned that she was on the ground, cooling off and vulnerable to attack by ants, Neil carefully coaxed the Duchess onto his finger and set her down on the top of a dried flower stem. I had imagined that butterfly watching meant dashing around wildly, chasing anything that moved, but this was quite the opposite. We watched and waited patiently. The sun reappeared, warming chemicals within her responded, her wings gradually opened and she was off.

'They are a rare and beautiful thing, don't you think?' Neil said.

Searching the other hollows, we found another female and were able to kneel down and get close-up views without disturbing her. Then, after a break for Thermos tea and a snack, we spotted two male Dukes spinning upwards in combat before returning to their territorial vantage points.

I had never spent time before specifically looking for butterflies; they had always been an added extra on days out. Rewarded by wonderful sightings in such a pleasant place – and, for a first-timer, a Duke of Burgundy wasn't a bad start – I could now appreciate the appeal of the interest. There is something delightfully old-fashioned about butterflies. They conjure up images of quaint cottage gardens, Edwardian vicars with nets and nature diaries, flower meadows, schoolboy collectors, woodland rides, summers past . . . Our affection for them is steeped in a sense of nostalgia, a yearning to return to a time when we had less of a stranglehold on our countryside. They take us back there, if only fleetingly.

The clouds soon rolled in and light rain began to fall. Heading back down the slope, we passed someone Neil recognised heading up to the site. He was a fellow butterfly watcher, and it looked like he had arrived too late.

'You should have been here an hour ago,' Neil teased.

Given the state of our beleaguered butterfly populations, 'you should have been here a century ago' might have been more appropriate. Long-term studies show that the vast majority of our species have declined in abundance and range, mainly due to the fragmentation and loss of suitable habitat. Drained of millions of butterflies, Britain is decidedly more anaemic than it once was. And given that they are regarded as indicators of the health of the countryside, this should act as a serious wake-up call. Except that we have got into the habit of slapping on the snooze alarm, rolling over and waiting for the next dire warning to prod us into consciousness. If good news is what it takes to motivate us, then one need look no further than Heyshott Escarpment, where successful endeavours to save the Dukes and Duchesses have brought back a little bit of colour to a corner of southern England.

It isn't easy being a dormouse in Britain.– although they always look remarkably relaxed about things. Suitable habitat for these endearing and dozy little creatures has halved since the early twentieth century, and their numbers have suffered a collapse akin to that of the England cricket team's middle batting order. The march of plantation forestry, the sprawl of urban development and the intensification of agriculture, which has turned fields into factory floors, has left fewer and fewer hospitable areas in which to make a home. But before we sharpen our pointy fingers of blame and have a go at politicians, farmers, developers and anyone else but ourselves about the environmental cost of modern living, consider the resources it takes to keep just one of us going.

Imagine piling up all you have ever owned on the street outside your front door: a few cars, perhaps, a couple of washing machines and dishwashers, throw on a fridge or two, toasters, kettles, old TVs and video recorders, bedspreads, music cassettes, light bulbs, several trunk-loads of clothes . . . Some items will have been recycled, shared or sold on, but most of us would be staring at quite a heap. Then start chucking on everything you've ever eaten. How many chickens? How many bowls of cereal? How many chocolate-chip cookies? Don't forget to add all the toiletries, water and fuel you've used so far over your lifetime. Then, when you're done – perhaps lobbing a few organic rice cakes and packets of Fairtrade coffee on the top to feel better about yourself – cover it all with an old tarpaulin and pretend it isn't your pile. Blame the neighbours.

Our growing demands and expanding population mean that much of our wildlife has been squeezed into the margins, dormice included. And these rare rodents hardly ask for a great deal to keep them going. They don't travel far, live in high densities, require large territories or have many young, and they spend more than half of their lives asleep – either hibernating through the winter or in a torpor during the day to save energy between nocturnal foraging trips. When the going gets tough, the dormouse gets napping. What they do require is woodland and scruffy hedgerows that can serve up a daily bellyful of nuts, pollen, insects, berries and fruit. Changes in the way we manage our countryside have scratched too many of these options off the menu in former strongholds, and dormice are now confined mostly to southern England and Wales.

However, all is not lost. In numerous woods, hands-on measures that adopt a more traditional approach to forestry, coppicing in particular, are helping to stem the precipitous decline of this much-loved mammal. Coppicing involves cutting trees down to a stump and harvesting the straight new stems that grow back, for fuel, charcoal production, fencing poles or building materials. Hacking trees rather than planting them might sound more like vandalism than conservation, but the injection of a little light and youthful vigour into woodland works wonders for flora and fauna. The dwindling of this centuries-old practice has had an effect on everything from orchids and nightingales to the Duke of Burgundy butterfly and the dormouse. Where it still takes place, or has been reintroduced, wildlife can flourish in the thickets and understorey – so long as hungry deer don't munch their way through the new growth first.

You don't get much better coppiced habitat than Briddlesford Woods on the Isle of Wight. This deer-free, four-hundred-year-old area of woodland is carefully managed for dormice by the People's Trust for Endangered Species (PTES), with plenty of other wildlife also reaping the benefits. If you're planning to see one, however, you would be wasting your time here, or anywhere else, simply wandering around hoping to get lucky. Dormice are small, secretive tree-dwellers that come out at night. To stand any chance you need to join an organised outing with licensed experts. So I signed up to help with a spring survey and arranged a day trip to the island.

I always leave far too much time to catch planes, and I bank on running for trains. But ferries throw me. I can never get the timing quite right. So I nearly missed the early crossing to Yarmouth despite leaving home at 4am – though for once I didn't get lost. All was going well for the first hour or two. Empty roads at dawn are a pleasure to drive on, apart from the fresh toll of roadkill that you come across, which this time included a small rabbit thrown into the kerb like a discarded glove, a couple of badgers and a hedgehog – something you see less often these days, dead or alive. Then, as the sky began to lighten, I noticed an array of signs in the distance. The A35 was closed, forcing me to follow a lengthy and time-consuming detour along winding 30mph lanes through villages, my stress levels ratcheting up, until eventually I got back on track and just managed to make it onto the boat as they prepared to winch up the gangway, peeling my fists from the steering wheel and climbing out of the car looking like I had just got off a fairground ride.

I had never visited the Isle of Wight before, and was looking forward to giving it a quick once-over. From the visitor guides on the boat it appeared that never a day went by without there being some kind of festival on the go. Miss one and you only had to wait half an hour for another to start up. Those advertised included the Chilli Festival, Walking Festival, Isle of Wight Festival, Jazz Festival, Real Ale Festival, Folk and Blues Festival, Garlic Festival, Literary Festival, festivals for vintage car enthusiasts, cyclists, sailors, gardeners and artists, as well as a Fringe Festival, which I presumed was for hairdressers.

The crossing was calm and scenic enough to soothe the nerves of any stressed motorist. After disembarking I headed for the centre of the island, noting that roadkill on this side of the Solent included red squirrels, and at Briddlesford Woods, close to Newport, I met up with Ian White, PTES's national dormouse officer. It wasn't long before other conservationists and reserve workers arrived, having travelled from all over to help with the survey and learn more about the protected species. Our small gathering had the makings of becoming the Isle's inaugural Dormouse Festival, though no one had thought to bring hotdogs and an acoustic guitar.

The woodland, designated a European Special Area of Conservation, was utterly stunning. A mix of old oaks and young saplings rose from a forest floor flooded with flowering bluebells. There were stands of coppiced hazel of differing ages and heights, tangles of bramble and honeysuckle, and a latticework of branches providing the kind of aerial walkways that ground-fearing dormice need to help them get about. They seldom cross open land and, as a rule, wouldn't be seen dead crossing a main road, unlike all the other species I had come across. Good tactic if you want to stay in one piece, but it does hinder their ability to colonise new areas.

One of the first clues that dormice are living in a wood is that they leave their clogs lying about – or rather, hazelnut shells that resemble clogs. The hazelnut is one of their favourite foods when preparing for hibernation. So much so that our native dormouse is also known as the hazel dormouse. But hazelnuts can be hard work. In order to get through the tough outer casing, they gnaw a circular hole in the broad end until they can reach the nut inside, prising it out and discarding the hollow shell on the forest floor. Not only does it resemble a wooden clog, but the chiselled hole has a telltale smooth inner rim.

There were plenty of hazelnut shells lying about, which could have been tackled by dormice. I found a couple of promising ones, but we didn't have time to sift through many. There were nest boxes to check – a mere 610 of them scattered throughout the forest. Everywhere I looked I could spot the numbered wooden boxes, fixed a metre or two above the ground and designed with the entrance hole facing the tree trunk. Dormice readily take to them, either for daytime dozing or to raise young in nests made of honeysuckle bark and dry grass. So, splitting into two groups in order to cover sections of the wood methodically, we set off to see how many were at home.

Of course there was never going to be one in the first few boxes. I had learnt from checks on reptile tins that animals know to hide in those furthest from wherever you start, just to make a game of it. And sure enough they were empty. Then again, the Bechstein's bat survey showed that some species don't want to play at all, although I felt certain that something as charming as a dormouse would be more cooperative.

Checking a dormouse box involves blocking the entrance with a piece of cloth, to prevent any inhabitants escaping, then carefully opening the lid and peering inside. You never know what you might find. Unexpected guests in Briddlesford boxes have included bats, pygmy shrews, various species of bird and a toad that had mastered the art of tree-climbing. We discovered a blue tit sitting on a clutch of a dozen light speckled eggs, which looked like Easter cake decorations, and an energetic wood mouse. Mice and shrews tend to breed fast and die young, with few making it through two winters, while dormice take a more leisurely approach to life and can reach the ripe old age of five. The larger, bolder edible dormouse, prized by Roman chefs and accidentally introduced to Britain in 1902, can live even longer. Found mainly in the Chilterns, it resembles a small grey squirrel and there are fears that it could become an invasive pest in the future because its range is starting to expand.

I was taking it in turns with others in our group to open the boxes, and my finds had amounted to an empty bird's nest and a few bits of moss. No one else was doing much better, though there was no sense that this was a contest or that we were racing the clock. The wood was an idyllic place to pass the time. I could hear the songs of robins, chiffchaffs and blackcaps, and between box checks Ian pointed out interesting plants with Latin names that sounded like Harry Potter spells: *Pulmonaria longifolia, Circaea lutetiana, Adoxa moschatellina, Gryffindor expelliarmus.*

'The dormouse helps to drive forward great woodland management,' Ian explained. 'What's good for them is good for so many other species.'

We crossed a muddy ditch, following paths through the bluebells, and I was handed the cloth to check a box. 'Prepare for the first dormouse of the day,' I joked with one of the other group members. With the hole bunged and latch unfastened, I raised the lid and could hardly believe it. I was right. Curled up at the bottom, fast asleep on a bed of leaves, was a dormouse.

I gently closed the lid and alerted Ian, trying my hardest to sound cool about it. He summoned the other group on his mobile phone, and soon everyone had gathered around the tree. Ian unhooked the box from the trunk, placed it in a large, clear bag and opened it again, carefully lifting out the sleeping resident and holding the golden brown ball of fur in the centre of his palm, to a universal sigh from those assembled.

'It's in a torpor, so we have about twenty minutes before it becomes active,' he said.

The dormouse, trying to conserve heat, was as tucked into itself as it was possible to be. With paws together, chin down and furry tail curled over the top, it could have been neatly gift-wrapped. Its tail was carefully unfurled to check the sex – it was a female – and she was weighed in the bag. She came out at nineteen grams, which Ian said was a good weight for late spring.

We then had the chance briefly to hold her, and she was passed into my cupped hand. For a moment she half opened her dark eyes, with a look that said 'five more minutes sleep, please', before shutting them again, unable to rouse herself.

When I handed her back, she left a lingering sensation of softness – not warmth, but softness – at the centre of my palm.

Dormice are undeniably adorable. From their chubby, hamster-like bodies, big eyes, rounded ears and long velvety tails to their dreamy, unhurried disposition, they have all the qualities needed to melt even the hardest heart. It had been well worth getting up at 3.15am to see one.

We returned her to the box, put everything back as it was and walked on, leaving her in peace. I wondered whether she would remember anything of what had happened when she woke up in the evening. Perhaps she would put it down to a surreal dream – too much cheese before bed.

'I never tire of them,' Ian told me. 'They are rare enough to be always exciting to see, but not too rare that they can't be seen. And they are fascinating animals, packing everything into six months of the year – putting on weight after hibernation, building nests, breeding and raising young all in a short space of time.'

A dozen or so boxes later we found another female, of similar weight, in a box near the woodland edge, and one more before lunch that was wide awake and made off up a hazel tree. It was surprising how quick they could be. Flexible ankles and sharp claws enable them to grip onto thin branches, using their tails to aid balance.

I had to leave the group in the afternoon and catch a return ferry. It would take another day to inspect all the boxes as part of the National Dormouse Monitoring Programme, and I later found out that a further nine dormice had been found, mostly females and all of a good weight, which was encouraging news.

I wasn't heading home but instead to Sussex to meet bat expert Daniel Whitby. A consultant ecologist and dedicated field worker familiar with all our Chiroptera rarities, I was hoping he could help in my search for a Bechstein's bat.

Dan had suggested I join a night survey he was carrying out for the National Trust in woods around Slindon, near Chichester, and we met at the old school in the village. His car was overflowing with equipment, but he managed to clear enough space for me to squeeze into the front seat before we drove off down narrow lanes into the forest to set up his traps.

It was back in his teenage years when Dan was first introduced to bats by an authority on the subject, who was monitoring radio-tagged species at an estate where his dad worked. 'I jumped at the chance to lend a hand,' he said. 'But I realised that, like many people, I couldn't name, let alone identify, a single species and I wanted to learn more.'

Years later, having accumulated a wealth of experience and knowledge, he is still learning, as is everyone interested in these mysterious creatures. We're still in the dark about many of their nocturnal habits – and you never know what you might discover. In 2012, Dan was surveying an area of the South Downs when he came across an overseas species called Geoffroy's bat. Whether it was paying a visit or part of a secret population already established here remains to be answered, but his find was the first ever recorded in Britain.

We parked deep in the forest, and I helped Dan set up one of his bat traps in a wide area between stands of mature oak. It was a 'harp trap', a twelve-foot-high rectangular frame threaded with taught vertical strings of transparent fishing line arranged in two parallel rows. The idea was that bats colliding with the harp strings fell unharmed into a soft pocket of material below. However, persuading canny flyers with excellent echolocation to flutter headlong into such a contraption is far from easy, even if you use neighbouring trees to conceal the metal uprights.

'Bats can detect a spider's web, so they can make out the lines of this trap – if they're looking,' Dan said. 'But by playing recordings of species' social calls you can elicit a territorial response and draw in curious bats.'

He flicked on the sound, a spinning reflector above giving the high chirping a sense of movement, and we moved to sites nearby to set up three more traps.

Once finished, we drove to another part of the wood and parked beside a National Trust cottage set back from the road. Gathered outside in the fading light were half a dozen people, all staring at the roof. A clandestine meeting of building inspectors? An outing of the local drainpipe and chimney society? No, these were bat enthusiasts helping with Dan's survey work and waiting for colonies roosting in the loft space to emerge.

At the rear of the house, common pipistrelles had begun to take flight, and a hand-held bat detector tuned to the right frequency meant I could hear their calls. Meanwhile at the front, seated side by side in the garden, a man and woman were enjoying an evening out watching a hole under the slates. Who said romance was dead?

'Can you hear that sound?' Dan asked me.

You could hardly miss it – a harsh squeaking as bats hidden out of view prepared for their night manoeuvres.

'Serotine bats,' he said.

One of our largest species, the serotine is found in southern England and makes extensive use of old buildings. It sounded as if the colony was arguing over who would be first to take the plunge. Then a single bat emerged from the hole and flew in a leisurely circle above the garden on broad wings. The bat detector picked up the low-frequency *tack-tack-tack* sound of its echolocation before it headed off over the trees. Others followed, and the two volunteers in the garden were kept busy counting. After about half an hour, we left them to it, by which time they had recorded thirty-eight.

Dan was concerned that the temperature was falling and that many bats might remain in their roosts to conserve energy, especially because hardly any insect prey could be seen in his car headlights as we headed back to check the traps. One surprise was an albino badger caught in the beams. It ran across the lane in front of us like a ghostly apparition, and disappeared into cover. Surely a good omen!

Two of the traps were empty, and in the third a couple of dark shapes could be seen clinging to the material beneath the harp strings. Reaching in, Dan picked them out, placing them in cloth drawstring bags, and after finding the final trap empty we joined the volunteers near the cottage to examine the catches. The first was one I had seen before: a brown long-eared bat, its huge ears uncurling as Dan placed it on his outstretched palm.

'People tend to lump bats together as just "bats", but it is like talking about birds as if they were all the same,' he said as the bat flew from his hand. 'They are as unique, in their own way, as swans, sparrowhawks and starlings.'

That said, distinguishing our small brown bats by sight, or sound, takes practice, even though we have just eighteen residents to get to grips with. In fact, some are so similar in appearance that they have been recognised as separate species only in recent years.

The second catch was a case in point. It was a pipistrelle bat, but not a common pipistrelle. Instead it was a lookalike soprano pipistrelle, which has a slightly paler face and a higher call, and was not separated from its virtually identical relative until the 1990s.

It had turned decidedly cool, and we spent the next few hours driving and walking between empty harp traps, changing the sound recordings every now and again. Bechstein's bats are late risers, only it didn't seem like they were going to bother getting up and about at all. It was past midnight and I was exhausted. Having been awake for almost twenty-four hours, I was close to slipping into a torpor myself.

Eventually we found a bat in a trap, and Dan looked it over intently before placing it in a cloth bag to examine back at the car. 'Interesting,' he muttered under his breath.

That's a good word to hear from any experienced naturalist.

'What is it?' I asked when he finally took it out and held it carefully between two fingers, scrutinising its features in the light of his head torch. It was a small individual, quite similar to the Natterer's bats I had seen in Dorset, with brown fur and neat little ears.

Using a magnifying glass, Dan peered at the cusps of its teeth, checking the heights relative to one another, then the pointed front part of its ear, called the tragus, before reaching into the car for a ruler and measuring its forearm. Finally, he held it out for me to look at. 'It's an Alcathoe bat,' he said.

Holy biodiversity, Batman! If telling the two pipistrelle bats apart in the hand is tricky, then this little critter, a mirror image of the whiskered bat and the Brandt's bat, is the toughest identification challenge of the lot. Experts in Europe had to resort to DNA analysis to confirm it originally as a distinct species, and it took until 2010 before scientists reported that they were living in our own backyard.

'This bat may be more common than the Bechstein's bat, but far fewer people have knowingly seen one,' Dan said.

He cupped his hands and breathed into them to warm the bat up, and once it was ready it took off, our torch beams tracking it until we lost it between the trees.

With the temperature now in single figures it was time to call it quits and help Dan pack away the harp traps. Once again, Bechstein's bat had eluded me, but the final encounter with the Alcathoe was certainly special. Perhaps the white badger had brought a little luck after all. And I was beginning to realise that my hunt for rarities was as much about the other animals I saw along the way. By refusing to give itself up too easily, bashful Bechstein's had introduced me to brown long-eared, Natterer's, noctule, common pipistrelle, soprano pipistrelle, serotine and Alcathoe bats – each and every one threatened by the loss of suitable habitat, roost sites and insect food. Initially I hadn't intended to come across any of these species, and now I was glad to have met them all.

# Six

After a busy few weeks of wildlife watching, I thought it best to devote a little free time to family life. Having disappeared at every opportunity, I was certain my wife and daughters would be struggling to cope – although, if they had missed me, they managed to disguise it remarkably well.

'Have you been away?' my eldest daughter asked, looking up momentarily from her laptop screen.

Her younger sister also did an admirable impression of indifference. 'Bad luck about the bat Dad. Can I have a lift into town?'

And my wife? 'Don't worry about us, we're fine,' she said, bravely. 'When are you off next?'

I told her I was driving to Suffolk the following weekend.

'Oh, that long. I mean, so soon?'

My daughters' schoolwork hadn't nosedived during the spring, despite an absence of fatherly guidance. In fact, their grades had unexpectedly improved. I was a bit disappointed, however, to see an 'F' in a column at the top of both their reports, until they pointed out that 'Gender' wasn't actually a subject.

Compensating for the guilt I felt over leaving them to, somehow, manage without me, I threw myself into domestic duties. I took over the shopping while my wife was at work and tried out new recipes with mixed success. The hallway smoke alarm turned into a kitchen timer as food under the grill was reduced to its carbon footprint, while all the desserts I attempted could have been named 'giant palava'. I vacuumed and dusted enthusiastically, until I ran out of steam, convinced the sort of people who might judge us by the cleanliness of our house were exactly the kind I wouldn't want to invite round anyway. Then I turned my attention to the garden, where slugs had munched their way through everything we had planted. Unwilling to resort to chemical warfare, my only hope was that they died of type 2 diabetes from overeating.

Yet however much I tried to keep busy and focus on the everyday, rare animals monopolised my thoughts. Mammals, fish, reptiles, amphibians, birds and insects roamed free inside my head. Delight at recent encounters was gradually eclipsed by rising anxiety over the logistics of finding other scarcities on my list, and I had a nagging sense that the year was ticking away, like a parking meter fed with too few coins. I needed to get on the road once more.

That weekend I said my goodbyes again, coaxing hugs out of my daughters. A fortnight's holiday stretched ahead of me, with a busy schedule of searches planned and a route that zigzagged from Devon to East Anglia, Cumbria and Scotland. The car tank was full of petrol and I had packed spare clothes, bedding, food, a camera, a map and other essentials to keep me going. I was ready for my next adventure.

***

The sun was sinking behind an arid plateau to the west. Burning heat and bleaching brightness ebbed away from the lowlands. Colours ripened in the softening light and the scent of flowers and fruit filled the humid air. Forest animals began to stir amid the lengthening shadows; in the distance, monkeys could be heard calling. It was April in south-west Africa, and high in the trees that lined a sandy riverbank a single bird was feeding. Obscured by foliage, it kept out of sight among the uppermost branches, racing nightfall as it gorged on figs and insects plucked from the leaves with its reddish bill. It was about the size of a thrush, with a vivid yellow body and black wings and tail. A bird of tropical brilliance. Just the sort of eye-catching species one would expect to chance across in the sub-Saharan woodlands and lush riverine valleys of this vast continent. And it had a glorious name to match: the golden oriole.

While other birds had gone to roost, this individual, a handsome male, was unable to settle. He had been restless for days. Some impulse was turning him on his perch, lifting his head skyward, tugging his wings open. He had spent the winter under an African sun, and now an internal clock told him the time was right to leave. The craving for motion was impossible to resist. He launched himself into the gathering gloom, climbing into cooler air before turning north. The place of his birth called him back. Guided by the stars, the movement of the sun and the earth's magnetic field, he would gradually work his way up the land mass, travelling by night to avoid the heat and raptors.

Over the days that followed, the golden oriole neared the Equator and passed over dense jungle. Mountainous thunderclouds forced him east, sapping precious energy, and by the time he had got back on course the forests lay behind him to the south. He flew over sprawling cities lit up like swirls of marine phosphorescence in the blackness, and he rested and fed by day among thorny trees before pressing on over the desert belt and the Mediterranean beyond. He was not alone. The sky was alive with the whirr of beating wings. A vast tide of migrating birds was sweeping into Europe, millions of them following aerial routes mapped out by their ancestors. Cuckoos, swallows, warblers, nightingales, pipits, flycatchers, whinchats, wagtails . . . all flooding onto the Continent to take advantage of long summer days and plentiful insect life in order to raise young.

As the golden oriole crossed Spain and France, many of these fellow migrants began draining from the sky, peppering the bushes and fields below and setting up territories. But he continued on, crossing the Channel by night, flying strong and true. And then the compulsion to keep heading north began to subside. It felt as if he had arrived. As dawn broke he saw poplar trees rising from low mist that covered a flat landscape, and he slipped down through the air and landed among their branches. He had reached Suffolk. It was mid-May, and despite having journeyed more than four thousand miles this tough little traveller found the strength to break into song.

One of the team at the RSPB's Lakenheath reserve was up early and walking along a path that ran beside the site's West Wood – a section of tall plantation poplars close to the river Little Ouse. A series of enchanting fluty sounds emanating from deep within the wood, pure and magical, like something from Tolkien's Elven valley of Rivendell, stopped him in his tracks. Although he was unable to see the shy singer hidden in the canopy, he knew the song well. Returning to his office, he posted a short message on the reserve's website. The first of the golden orioles was back.

An estimated sixteen million birds funnel into Britain from Africa every year, but very few golden orioles are among them. The number that breed here, weaving hammock-shaped nests in the forked branches of lofty trees, has plummeted from a high of more than forty pairs in the 1980s to one or two pairs at best today. For some unknown reason they are wedded to poplar plantations in East Anglia, and the felling and deterioration of favoured stands, originally planted as timber for the matchstick trade, may have contributed to their decline. The future for our popular poplar populations of orioles hangs in the balance.

Lakenheath RSPB reserve is one of the few places where golden orioles can regularly be seen, even though their breeding success has become a hit-and-miss affair here, as elsewhere. I had read the website message, and pulled up in the car park a few days later brimming with anticipation. It was a sunny Sunday with plenty of visitors about, and I made my way past one of the main reed beds to West Wood, where I joined a few dedicated birdwatchers strung out along the path, prospecting for ornithological gold. They were happy to swap information and anecdotes, but none had glimpsed an oriole that day.

A narrow channel of water prevents access to the wood, so you have to stand on the outside looking in, patiently scanning the shady corridors between regimented rows of straight poplar trunks and the swaying tops that meet high above. Despite the striking, hazard-sign contrast of their yellow and black plumage, golden orioles are notoriously difficult to see among the rattling round green leaves, and a good ear is as valuable as a keen eye when looking for them.

The male, if present at all, was refusing to utter a single note. Still, there was a fantastic array of other birdlife to divert my attention, including sedge, Cetti's and reed warblers, a visiting red-footed falcon from eastern Europe, cuckoos, reed buntings, whitethroats, hobbies hunting dragonflies over the reeds, swifts, marsh harriers, scarce bitterns and cranes, and bearded tits. Not bad considering this whole area had once been a carrot farm.

After a few hours of staring at trees, my senses began playing tricks on me: a blackbird singing, a patch of light-coloured lichen, a leaf twisting in the wind . . . all orioles, at least for a second. Afternoon turned to evening and the crowds dispersed. It seemed a lost cause, but I continued my watch. I had enjoyed getting to know the sounds of this place, from the booming of bitterns to the scratchy chattering of warblers, and a new noise startled me – a 'plop!' from the other side of the ditch. A water vole had dived from the bank and was motoring across the channel like a clockwork toy, nose up, leaving a trail of ripples behind it. It was the first time I had ever seen one. A little treat served up before I had to leave.

I slept in my car in a village nearby and returned to the reserve at 4.30am. Dawn is the best time to hear a golden oriole singing and I had the place to myself. Roe deer bounded away through the long grass, a grasshopper warbler was reeling and, bizarrely, I spotted a stoat swimming across a short stretch of water, perhaps hunting water voles. I took up my position beside West Wood and waited. It was drizzling slightly – not exactly ideal weather, but I was still optimistic.

By 9.30am that confidence had waned. Reserve staff passing along the path nodded sympathetically and wished me good luck. They had no new intelligence. My vigil continued. Then, at about 11am, two birdwatchers with scopes slung over their shoulders came along the track.

'We've just heard it,' one of them said calmly.

'Really? Where?'

'Over in Trial Wood.'

I loved the fact that the subject of the conversation remained unspoken. They had rightly guessed what I was searching for, only now it seemed I had been watching the wrong patch of poplars. I thanked them, grabbed my belongings and ran to the neighbouring triangle of trees.

Unsure exactly where the best vantage point was, I circled the plantation and repeatedly paced a track by the deepest section, until a burst of birdsong took me by surprise. A sweet cascade of liquid whistles, of the kind one might hear echoing through equatorial rainforest, filled the woodland, pushing all other noise aside. This was it, a male golden oriole proclaiming himself with a refrain straight from the heart of Africa. You could almost taste the nectarous tropical notes as they fell from the canopy. It was a delicious sound. Yet while I was practically standing under the tree where he was singing, I simply couldn't spot him.

Frustratingly he fell silent and I doubled back to the main path, where I met a couple of birdwatchers also hoping for a glimpse. After a while it seemed like the chance had gone. Then something caught our eye – a bird flying from the corner of West Wood, close to where I had been stationed earlier. It flew in a straight line towards us and, as it passed over and disappeared into Trial Wood, we got a good look at its underside. Its chest was pale and streaky with a hint of greenish yellow nearing the tail, and it had a thick bill, not overlarge. We all knew what we had seen, and our bird guidebooks confirmed our suspicions: it was a female golden oriole. An incredible sight, because generally they evade detection, lacking the vibrant colour and voice of the male.

After fourteen hours I had been rewarded with a song, and a view, of a bird close to extinction as a breeding species in Britain. And the icing on the cake came on returning to the reserve visitor centre, where a site map on a metal board was dotted with magnets showing the day's sightings. I was able to pick up the unused 'golden oriole' lettering and place it proudly over the western end of Trial Wood – just in case any hopeless rarity spotter was having trouble finding one.

I spent the night at a sixteenth-century hotel in nearby Downham Market, in a room resting on such bowed old beams that every piece of furniture had wedges underneath to keep it horizontal, or perhaps to prevent it from sliding down the sloping floor. It was the first incline I had seen while travelling in this part of Britain, and I wouldn't see anything else resembling a gradient for a day or so because I was heading further into the heart of the Fens. The area's vast arable plains under huge skies took some getting used to, but were certainly easy to navigate. In Devon, ancient hedge-lined lanes are set deep in the undulating land. Eroded by rain and traffic down the years, they frequently lie lower than surrounding green fields, so it seems as if you are tunnelling through the terrain. By contrast, the marshland that once covered this part of East Anglia has been drained over centuries, and the exposed expanses of wind-blown, sun-baked soil have shrunk to such an extent that roads stand proud of the flat landscape. It was like driving across an ironed map. Even I couldn't get lost.

I was going fishing once again, and the next day I made for Outwell, north of Ely, and pulled up outside Peter Carter's shop opposite the post office. You don't get many shops like Pete's any more. Chaotic and characterful, its windows were crammed with paintings, books, nets, curiosities from the area's past and an assortment of different signs: 'Mole trapping', 'Rabbit nets', 'Outboards serviced', 'Rural books', 'Basket willow sold here', 'Wood carving', and the reason I had come: 'Fresh Fenland eels'. Pete is the last traditional eel fisherman in this part of the world. He follows in the muddy footsteps of forebears stretching back five hundred years. At the rear of his shop, between bundles of willow, dusty old reels and woodwork tools, stand a pair of boots and a multi-pronged fishing spear that once belonged to his great-great-great grandfather.

'Generations have passed down the secrets, and I was taught by my grandfather,' he said as I joined him in his Land Rover. 'But that knowledge is being lost as children today don't want to do it. In many ways I don't blame them – there's not a lot of money in it.'

Aged in his forties, Pete is no white-whiskered old-timer, but he does look and sound every bit the eel fisherman, with his weather-beaten clothes, unhurried manner and local turns of phrase.

We parked on the outskirts of the village by Well Creek, a narrow natural watercourse that connects the Nene and Great Ouse rivers, and Pete unloaded his gear into a small wooden boat tied up beneath the willows. I was wearing two jumpers, thick waterproof overtrousers and a woolly hat, but was still cold.

'When the willow comes into bud, the eels come out the mud,' Pete recounted in his leisurely Norfolk accent. 'That's what they say. But it's been so cold I reckon the buggers went back in again. It's been really bad so far this year.'

It is not just the weather that makes it tough for him to earn a living. Numbers have fallen so catastrophically that, like the common skate, the European eel is classified as critically endangered. This is a staggering state of affairs given how abundant they once were. In times gone by, Fenland villages paid their church taxes in eels, by the thousand, while just down the road on the 'Isle of Eels', glorious Ely Cathedral was built by trading barrow-loads of the fish for Barnack stone. The city's namesake is even celebrated with an annual festival where you can try your hand at eel throwing – toy replicas now provided.

Our wriggly times of plenty are long gone. Eel populations have plummeted ninety-five per cent, with factors thought to include large-scale commercial fishing operations across Europe, changing ocean conditions, the spread of parasites and the replacement of leaky wooden weirs and dams with metal and concrete constructions, which block the passage of eels in waterways or dice them up in hydroelectric turbines. Whatever the causes, Pete has seen catches dwindle over his lifetime. One of a few people licensed to trap eels, he uses old-fashioned baited baskets woven from willow, as well as small net traps designed to funnel fish through one-way openings into cylindrical chambers of mesh. The limited numbers he takes are sold to restaurants, and he also helps with tagging schemes and shares his knowledge with scientists trying to conserve the species.

'I used to get more than a hundred eels in a night, my best was 180, and now a couple of dozen is good,' he said. 'But this year's been awful. I got none in the nets the last time I put them out.'

I took a seat on a strut at the front of his boat, placing my feet in the puddles covering the wooden boards – 'The floors are a bit dodgy,' he warned me – and we motored slowly down the winding creek, which ran alongside the two-lane Downham Road and was about the same width. The river was higher than the arable land that surrounded it. In any other place that would defy logic. And, as if to confuse things even more, an aqueduct carried us over another river – a large, straight drainage channel that ran like a wet, grey motorway from one low horizon to the next.

Eventually Pete cut the engine and used a pole to punt us along the edge of the reeds that fringed the sluggish brown water. It was here that he set his net traps, away from pleasure boats and out of sight beneath the surface. Only the tips of willow-frond markers that broke the surface gave their position away.

I should have told him not to bother with the first net. Just like dormouse boxes, reptile tins and bat traps, it would likely as not be a lost cause. And, sure enough, I was right. The line of netting a few metres long had guided only a couple of small perch into the tunnels of mesh positioned at either end. The second and third net traps were also eel-free.

'Well, I've paid my dues,' Pete sighed as he lowered the anchor and rolled a cigarette. 'A silver coin thrown in the water for luck when I set my nets. Just a fivepence though – I ain't rich enough for real silver.'

He sat smoking, gazing out over the rectangular fields of yellow, dark brown and green laid out like carpet tiles as far as the eye could see, and talked about his childhood spent messing about outside when he should have been studying.

'I've spent my whole life on the water,' he said. 'I don't know much else. But for me this is the place to be.'

I envied his working week outdoors – though not those disheartening days trudging home empty-handed. Eels are certainly unpredictable and mysterious creatures to get tangled up with. They were once believed to originate from mud, or dewdrops, thatch or horse hairs. Today our best educated guess is that they breed in the depths of the mid-Atlantic's Sargasso Sea. Their tiny larvae, shaped like willow leaves, then drift for thousands of miles on the ocean currents before arriving a couple of years later around our coasts as translucent glass eels. Heading up our estuaries in spring and into our rivers, they become darker elvers and seek out suitable stretches of water in which to settle, crossing land if need be and feeding on anything going.

The brown adults with yellow bellies – 'barley eels', as Pete knows them, and the resident fish we were after – can live here for more than twenty years before they reach sexual maturity, whereupon they lighten in appearance and swim back out to sea on moonless autumn nights as 'silver eels', to begin the long journey back to their spawning grounds. It's quite the opposite life cycle to that of Atlantic salmon, which migrate upstream to lay their eggs.

Pete pulled up the anchor and pushed us on into the cold wind. The next 'end' was empty, but his dues were repaid with the net that followed. 'There you go,' he said, lifting it up beside the boat to reveal a small brown eel writhing inside the tubular mesh. 'A barley eel. Not that big, but it's a start.'

Handling such a slippery catch is no easy matter. Their velvet-smooth sides, slick with slime, glide through your grasp. Pete used the net to grip the sinuous length of liquid muscle and fed it like oiled cable into a bucket half filled with water. It circled the base and snaked up the sides, displaying its glossy head with small eyes.

'I think they're a beautiful colour,' he said. 'But a lot of people don't like them at all.'

Their appearance certainly doesn't endear them to your average shopper. Once a staple part of our diet, they have fallen out of favour. However, traditional East End jellied eels, brimming with trendy cockney TV chef appeal, are now finding their way onto our supermarket shelves. Pete prefers his fried in butter (though perhaps this isn't the place to be giving recipes for critically endangered animals). Either way, an appetite for eels remains high overseas, particularly in Asia, and because they can't be bred in captivity, glass eels are harvested in large quantities and grown on in fish farms to meet demand. The idea is that commercial operations in Europe pay their dues by helping to replenish depleted rivers. But with falling wild stocks pushing up prices, not everyone is prepared to let sufficient numbers slip through their fingers.

The following net produced a far bigger eel, a 'lively bugger' at least a couple of feet long, and Pete looked relieved. Perhaps the slow start to the season was behind him. Checks further along the river added two more of a similar size to the bucket, along with a smaller one, before it was time to head back. They were mesmerising to watch, circling in the water, like some primitive missing link.

I left Pete in his shop, among the willow poles and traps, and thanked him for a fascinating and relaxing morning. What better way to unwind than punting along a river with someone who doesn't own a watch? Scarce animals aside, 'eel man Pete' was certainly a rare and endangered kind of person, endeavouring against the odds to keep alive not only an unusual family occupation but also an ancient fenland tradition. For how much longer remains to be seen, given how rapidly his quarry is vanishing.

Stress returned fairly swiftly when I hit heavy traffic on the M6. Squeezing into bumper-to-bumper queues, I crawled north towards Cumbria, lurching between first and second gear and cursing every wasted minute. Surely this many people couldn't also be going to look for natterjack toads? A charismatic species they may be, and late spring was the prime time of year to find them, but did tens of thousands of amphibian enthusiasts really have to pick the same day as me for an outing?

The dense traffic began to pick up speed by Preston, thinned out past Lancaster, was sparse as I joined the A590 to Barrow-in-Furness and all but gone as I turned off onto the road to Roanhead. I arrived at Sandscale Haws National Nature Reserve in the early evening and found I had the car park to myself. Where the crowds of fellow toad-spotters had got to I had no idea. The weather certainly seemed favourable enough.

I hadn't needed to rush. You can never be too late for natterjacks as they are mostly active at night. By day individuals hide in burrows – toad in the hole, if you will – then as the sun goes down they emerge to feed and breed.

Compared to our widespread common toad, the natterjack toad is extremely rare, living in Britain at the northern edge of its European range and confined to a few dozen heathland and coastal sites dotted around Cumbria and Merseyside, the Solway Firth, Norfolk and Hampshire, among other places. As amphibians go, natterjacks have pretty much torn up the rule book. For a start, they are such weak swimmers, with short hind legs and tough, non-breathable skin, that they can actually drown. They also run, rather than hop, sprinting on squat limbs through short, tussocky grass after beetles and other insect prey. Instead of damp surroundings, they favour hot, dry habitat, which is why sand dunes – perhaps the last setting one would normally expect to find toads – are good places to look for them. And the grassy banks of Sandscale Haws on the edge of the Duddon estuary boast the highest concentration of natterjacks in the country.

The view from the National Trust car park north-west across the wide bay was spectacular. Lake District peaks rose in the distance: Scafell Pike, Crinkle Crags, Coniston Old Man. To the west stood a line of wind turbines, not all turning despite the breeze – a motionless wind turbine, frozen mid-rotation like a stopped clock, is a strangely unsettling sight. On the beach a few dog walkers were catching the last of the light, some following paths up through the dunes, and I could make out the piping sound of oystercatchers above the gentle sighing of the sea.

Opposite the car park a simple sign, a toad silhouette with an arrow, pointed the way down a boardwalk, and I followed it to two ponds side by side just above the beach. One was circular and the other rectangular, both fenced off to keep people and dogs out, and a freshwater stream ran between them to the shoreline. These were not lush, deep pools topped with lily pads, but exposed sandy scrapes, a foot or two deep and virtually devoid of vegetation. And that's just what suits natterjack toads: shallow, sunny ponds that are prone to drying out. Their tadpoles develop fast in the warm water, outcompeting rival species for food and clambering out as toadlets within a few weeks. Aquatic predators, such as fish and dragonfly larvae, can never get firmly established in these ephemeral pools, but toad offspring also pay the price if the water evaporates too soon, which is why natterjacks spread the risk of losses by having a breeding season that lasts from April until July.

I leant on the railing and peered into the circular pool. Nothing there. I was too early. But I did notice a string of spawn lying on the bottom. This discarded necklace of tiny black pearls was the spawn of a natterjack toad, containing thousands of eggs. Precious potential strewn in the shallows, with no effort made to conceal it.

The rectangular pool contained three more strings deposited in double tramlines at one end. At the other, a length of spawn a few days old had broken apart and the black eggs and tiny curled tadpoles were piled amid the algae like a heap of punctuation.

A couple of visitors joined me at the water's edge and we got talking about toads, warts and all, and a little while later one of the local residents walking her dog also stopped for a chat. Wildlife brings people together like that.

'I can hear them croaking through my open window at night,' she said. 'It's a wonderful sound.'

Male natterjack toads are famous for their deafening mating call that can be heard more than a mile away. Amplified by an inflatable vocal sac, the croaking lures females to the breeding ponds, and I couldn't wait to hear it – not least because it offered the only chance of spotting one of these shy and well camouflaged animals after dark.

The beach and reserve were soon deserted, the ragged edge of Lakeland fells faded into black, and I sat on a dune in the moonlight looking out at the twinkling street lamps of Haverigg and Millom on the other side of the bay. It was a little unnerving being out alone in such an unfamiliar place, and one can't help but feel a bit suspect lurking by a beach at night. Not easy to explain should I be challenged by a passing police patrol.

'Toad spotting you say?'

'Yes, that's right, I . . . '

'Very convenient, sir.'

'Well it is a good site, though I live . . . '

'And you wouldn't happen to know anything about that pile of bodies behind you?'

'What? No, I . . . Bodies?'

'So you do know about them.'

'No, you just said, I was just . . . '

'All croaked it accidentally did they? Reckon it's time we headed to the station for a natter.'

My imagination was getting carried away. I flicked on my torch, drank a little Thermos tea, checked for bodies, had a snack and looked in the pools. Still empty.

It had gone 11.30pm when the silence was broken by a sound that was as comical as it was surprising. A toad began calling from the area of the rectangular pool. Similar to the noise those wooden toy frogs make when you run a stick over their serrated backs, and louder than anything I had heard emanating from a pond in Britain before, the croaking started slowly, then rose into a continuous rasping more suited to an Amazonian swamp than a headland in north-west England. It was so ridiculous I laughed out loud.

I climbed down the dune and tried to locate the source. It seemed to be coming from under the boardwalk out of view and I half expected to find hidden speakers playing a recording on a loop. As I approached it fell silent, so I kept still and waited until, a couple of minutes later, it started up again, this time a short distance in front of me. I switched on my torch and slowly played it over the pool, eventually spotting a shape at the edge: a toad facing outwards, throat sac ballooned, shredding the night air with his vibrato call. As I crept nearer he momentarily ducked below the surface before rising and standing proud of the water, grinding out his song. Lighter in colour than a common toad, his bumpy brown skin was blotched with green and flecked with red, and a thin yellow stripe ran down the centre of his back – a key distinguishing feature. I had found myself a natterjack toad, and one doing what they do best: making a racket. A second joined in from the circular pool, one further off revved up, then a fourth, until it felt as if the whole reserve was reverberating with the surreal sound of serenading toads.

It had been a long day, and after a while I left them to it and headed back to the car park. One advantage of sleeping in your car is that you can stay up, or get up, as late or early as you want, without having to abide by B&B and hotel rules. Falling asleep in my car listening to the toads was a nice idea, but I felt a bit uneasy about the remote location and drove into town. Parking in a well-lit area by the hospital, I managed a few hours' sleep before returning to the reserve at dawn.

The rising sun had thrown a warm blanket of orange light over the distant fells, and the air was still and silent. I wandered down boardwalks that were wet with dew and checked the ponds. The toads were gone, but a couple more strings of spawn had been added during the night.

I had arranged to meet up early with National Trust ranger Neil Forbes, and when he arrived he made a note of the new spawn. Female natterjacks produce one string a year, so by surveying pools conservationists can estimate how many toads they have on their patch. Sandscale Haws supports more than a thousand breeding adults, and they not only use the two main ponds, which are drained and cleared of vegetation every winter, but also other shallow stretches of water dotted around the area.

'For its size, this reserve has the greatest density of natterjack toads in Britain,' Neil said. 'It's a really important native population, especially given how many suitable heathland and dune-grassland sites have been lost.'

Some of his guided walks attract sixty visitors hoping to see and hear natterjacks – not quite enough to bring the M6 to a standstill, but impressive turnouts nonetheless. Amphibians are the most threatened group of vertebrate animals in the world, so it's encouraging to know that here, at least, people care enough to make the effort.

'To be out with so many of something so rare,' Neil added, 'and to be surrounded by a full chorus at night, really is a powerful experience.'

Even the few I heard had been remarkable enough. The sound would stay with me, and I knew it was something I would want to hear again in the future.

Why does the croaking of a toad, or the song of a golden oriole, the muscular feel of an eel, the sight of a sleeping dormouse, the elegance of a smooth snake or the delicate beauty of a Duke of Burgundy butterfly move us? Our responses may be learnt, guided by prevailing attitudes towards certain animals and outdoor experiences. But perhaps there is something deeper still, some innate affinity with other beating hearts of differing shapes and sizes, some sense of intimacy that disregards immense evolutionary divides. An appreciation of nature may indeed be part of our nature. And the connections we forge with the living world are now acknowledged to be essential to our well-being. Flowers by a hospital bed and tapes of birdsong have even been shown to help aid the recovery of patients. While I'm not sure about the health benefits of listening to natterjack toads, I certainly felt the better for having heard them. At the very least it cancelled out any psychological damage caused by the stressful motorway drive to get there.

Just as well, because I had another day on the road ahead of me. A similarly rare sound called me even further north, across the border into Scotland. It was comparable with a croaking toad, but with a harder edge and a monotonous chafing rhythm: *krrek-krrek, krrek-krrek, krrek-krrek*. Although I had heard it before, years earlier, I longed to listen to it once again. Surprising, perhaps, given its lack of musicality. Then again, just like a chorus of natterjacks, repetitive and simple sounds can be among the most evocative: the buzzing of bees, a woodpecker hammering on a branch, a cat purring, a cuckoo's two-note refrain. *Krrek-krrek, krrek-krrek, krrek-krrek* was up there with the best of them.

From Cumbria I drove to Oban on the west coast of Scotland. I was back where I had started, only this time there was no skate-fishing crew waiting or whale in the bay to welcome me. I was catching a morning ferry to the small island of Coll, one of the Inner Hebrides, and spent the night in a cheap B&B before turning up at the terminal. The wind was nearing gale force and the Caledonian MacBrayne service was on amber alert, with cancellation possible. However, I was glad to see the brute of a boat being loaded up and, after a short delay, passengers were allowed to board and we headed out of the harbour. 'It might be tricky getting alongside at Coll,' the captain warned us over the intercom, 'but we'll give it a go.'

Out past the skate-fishing grounds, and through the sheltered channel between the Isle of Mull and the Ardnamurchan peninsula, was steady going. Only when the ferry began to cross open water did the wind make its strength known. Large waves sent spray over the decks and windows, and the turbines juddered between swells as their grip slipped in the rolling sea.

There was a minor emergency on the lounge deck close to where I was sitting. An elderly couple a little way behind me got into difficulty as the ship fought its way through the Atlantic breakers. I could see the man looking perplexed, and the woman beside him tugging at his arm and asking: 'What is it?'

'Um, I'm not sure,' he replied, staring out of the window into the distance. 'I . . . I can't see properly.'

'Perhaps we should ask someone for help?' she whispered.

After a minute or two I could sit by and let them suffer no longer. I stood up and rushed to their aid.

'Excuse me, but I couldn't help overhearing your problem and thought I might be of some assistance,' I said gallantly. 'It's a gannet.'

'Ah, we wondered whether that was it,' he said, looking grateful.

'Oh, a gannet,' she added, glancing through the window as it flew parallel to the boat a little way off. 'Big, isn't it?'

'Yes, it's our largest seabird,' I added, before returning to my seat.

Cometh the hour, cometh the man. I shudder to think what basic identification errors might have been made if I hadn't stepped in. I'm only surprised the captain didn't put out an appeal on the public address system: *'Ladies and gentlemen, do not be alarmed, but we have an urgent ornithological situation on board. Can I request anyone with birdwatching experience meet at the Deck C rendezvous point immediately. I repeat: all birdwatchers to Deck C.'*

After an hour and a half the island of Coll came into view: a low, rough-edged land mass of heather moorland and grassy fields dotted with knuckles of grey rock. A few scattered houses and a simple jetty jutting into the foaming sea were the only signs of human habitation.

The captain had been right to foresee problems trying to dock. Buffeted by the gale, the ferry struggled to get close to the quay. Eventually a rope was thrown to the harbourmaster, who looped it over a mooring bollard, and with engines on full throttle it took several attempts to swing the boat broadside against the full force of the wind before it could be tied into place.

I felt quite relieved to step ashore, even into such a bitter blast, and soon warmed up as I climbed the mile-long road that ran around a bay to the main settlement of Arinagour. This little fishing village, home to a large proportion of the island's two hundred-strong population, consists of tough terraced dwellings and stone houses skirting the shoreline. Lobster pots were piled high in gaps between the properties, and a few fishing vessels had been hauled from the water, up over stranded masses of seaweed and onto patches of grass kept short by free-ranging sheep, which eyed me through coin-slot pupils as I passed.

I was booked in at the newly built Coll Bunkhouse, a bright and clean hostel on the edge of the village, and I dumped my rucksack on a top bunk, grabbed what I needed for the day and set off on foot.

Coll is shaped roughly like a fish, with Arinagour halfway down the body. At twelve miles long and three wide, it is not exactly a large island to get around, and I followed the only road south towards the tail end, which lies close to the neighbouring island of Tiree. Passing through one of the few copses of trees clinging to this windswept terrain, the road ran across rough pasture before climbing over heather moorland, where small lochs sat amid sulking expanses of boggy ground. It reminded me of Dartmoor, but the bones of this place are far, far older. The hard metamorphic rock that underlies the land and punctures the peaty soil is among the most ancient in the world. Created by immense heat and pressure deep within the earth's crust, and known as Lewisian gneiss, after the Hebridean island of Lewis, its origins date back nearly three thousand million years, spanning two thirds of the entire history of the Earth.

Of course plenty has happened to the rock in the intervening eons since it was created – including a few global adventures before arriving at its current location. Half a billion years ago you would have had to search south of the Equator to find the segment of land that was to become Scotland, while England was at that time embedded in a separate continent several thousand miles away near the South Pole – a Scottish nationalist's dream.

Around four hundred million years ago, the two 'countries' were brought together as their distinct land masses collided, then were swallowed up within the monster continent of Pangaea. As this giant drifted north and tore apart, Britain was baked dry under a desert sun, submerged in tropical seas, shaken by earthquakes, scalded by volcanoes, thrust upwards, ground down, stalked by dinosaurs and, having come to rest at its present geographical position, frozen over and scoured by glaciers. Finally, as the last ice sheets began melting about ten thousand years ago, sea levels rose to create the British Isles we know today. Our land link with Ireland and Europe was severed, and coastal valleys fringing western mainland Scotland, as elsewhere, were flooded. The maritime islands of the Hebrides found their places on the new map of the world.

Coll was a remote and challenging outcrop to inhabit, no doubt about it, poking its head from the sea into whatever the passing weather and dependent on links with the mainland for supplies and visitor income. Cut off and laid bare before the elements, this open and honest landscape of wire-fenced fields and rocky moors appeared to have nothing to hide. It felt good to be here. The clean air knocked the tangles from my brain and the leisurely hike stretched the knots from my limbs.

I had read that it is safe to hitchhike, and decided to give it a go. It's not something I had done in a long while, but this was just the kind of isolated spot where it would be rude not to trust in the generosity of others. So I slowed my pace when I heard a car approaching on the single-track road behind me and lifted my arm.

A small, battered vehicle stopped and the driver wound down his window. 'Did you want a lift?' he asked with a friendly grin.

'If that's okay.'

'You were a bit vague, that's all.'

'I guess I wasn't sure exactly how much I wanted one,' I smiled.

'Enjoying the walk then? Where are you heading?'

'The RSPB reserve.'

'It's a way to go, but hop in and I can drive you to the end of this road.'

A woman and child were also in the car, along with a dog, gas canisters and boxes, and there was hardly space for another passenger. However they kindly made room and took me across the moor, teasing me about the need to be more decisive when thumbing a lift, and dropped me off at a junction. This was one of the flattest areas of the island, and the road I followed, bordered by low stone walls on either side, made its way along the edge of a broad valley that linked sandy beaches to the north and south. The fertile plain had been divided into arable fields, while sheep and cattle grazed the higher rough pasture. I had been glad of the lift, but walking helps regain that sense of space you lose by traversing the countryside at speed. I could gradually learn to slow down in a setting like this, if I only had more time.

One thing I noticed were corners and corridors of thick vegetation in the fields, fenced off from grazing livestock. These patches of nettles, sedge, flag iris and cow parsley provided a spring refuge for the scarce and elusive species I had come to see, and despite the cold wind I paused every so often along the road to watch and listen. I was looking for a bird – a ground-hugging, cover-loving bird that does its level best to avoid being seen. On paper it doesn't seem much to get excited about: a streaky grey-brown relative of the moorhen which resembles a slightly stretched quail and has an unrelenting mechanical call that barely qualifies as birdsong. Yet it has become not only a conservation cause célèbre, but also a favourite among birdwatchers. The species? The corncrake.

Like the golden oriole, the corncrake spends the winter in central and southern Africa, migrating here to breed during our warmer months in grasslands and meadows. And it is even more secretive. As soon as corncrakes touch down on British soil in April they go into hiding – sneaking about in whatever dense vegetation will screen them from predators, and only getting airborne as a very last resort when disturbed. If they kept quiet we wouldn't know they existed. Then again, they probably wouldn't exist, given how impractical it would be to find a mate concealed silently in the long grass. So they call out to each other. Heard and not seen is their motto, and males stake out territories and attract females with their repetitive *krrek-krrek* sound, especially after dark, when they utter it almost continuously with head thrown back and beak flipping open. It is calculated that during the breeding season an individual can repeat this ratchet double-note up to a million times.

The call of the corncrake was at one time the backing track to summers across rural Britain. Farming created the insect-rich hay meadows in which the birds raised their young, and they thrived right across the country. So much so that, in the 1860s, Mrs Beeton recommended using a generous three or four in her recipe for roasted 'land rail', as they were then known. Towards the end of the nineteenth century, however, all that began to change. As harvesting methods mechanised and grass was reaped earlier in the year, corncrakes were literally mown down with the crop. It was a case of live by the sward, die by the sward. Cutters circling a field herded these panicky agoraphobes into the centre, where they hid in the diminishing cover waiting for danger to pass, unwilling to take to the wing or run for it across open ground. The result was that adults and chicks were destroyed, and a population that numbered in the tens of thousands drained from the countryside until, by the 1990s, they were as good as extinct in England and Wales and confined largely to the crofting fields of the Western Isles of Scotland. In 1993, calling birds numbered just 480.

Today they are being helped back from the brink, and that low figure has more than doubled, but it is slow progress. Steps include providing well-vegetated corners in which corncrakes can hide before the meadows have grown up; cutting hay later in the summer when chicks have left the nest; and mowing fields from the middle outwards to drive the birds into the margins and out of harm's way.

The traditionally managed hay fields of the Hebrides remain their stronghold, and Coll has over one hundred calling corncrakes, with more than half on the RSPB reserve. As I neared the site and the sun broke through the clouds, I felt increasingly confident that I would get to hear one of these globally threatened birds, even if seeing one was quite another matter.

The reserve at Totronald is gratifyingly inconspicuous. No large entrance signs greet you, and there isn't a paved car park, gift shop, café or toilets, just whitewashed stables housing an information display next to the warden's cottage and more than twelve hundred hectares of farmland, moorland and dunes to explore. Not that I needed to go far in my search for the area's star species. I had been right to be confident – the moment I arrived I heard a corncrake. It was just over the road from the stables, calling in an overgrown area. *Krrek-krrek, krrek-krrek.* And then another piped up from dense vegetation beside the buildings. Walking up the front path, I tried to pinpoint the exact direction, and as I rounded the side of an old barn there it was, standing out in the open in a patch of low grass, as easy as that. A peculiar little bird at first glance, but attractive in an understated way, with a slender body, pinkish pointed bill, chestnut-coloured wings and light brown plumage marked with black, gold and grey. It seemed as astonished to see me as I was to clap eyes on it, and off it ran into a thicket, weaving between the nettle stems and vanishing.

It may not be the most straightforward reserve to visit, but when it comes to corncrakes Totronald certainly delivers. I couldn't have asked for more in my first five minutes. No great search, no trekking through endless fields, no patient waiting, no maddening frustration trying to pinpoint calling birds. The celebrated master of hide and seek had given itself up in an instant and, retreating into the stable block to get out of the wind, I could still hear it through the window, craking away like an unanswered phone.

Once I had warmed up I went for a walk to the dunes overlooking the sea and strolled back across open fields of lime-rich soil lightened by shell sand and known as machair. Lapwings rose up ahead of me, their kazoo calls as un-bird-like as those of the corncrake, one of which could be heard a little further off. The pair of them sounded as if they were auditioning for an eighties synth band. A flock of finches – twites and redpolls – joined in, and I spotted a redshank on a fence post, a wheatear, two snipe, a singing skylark and a pair of rock doves, pure-bred ancestors of the street pigeon. This reserve was an I-spy paradise for birdwatchers. No meadow pipits? Ah, there they were. Reassuring to find a habitat and its species so neatly matched. It was like wandering through a picture spread in a book on British wildlife. Everything in its rightful place.

Returning to the stables, I bumped into reserve warden Ben Jones, and we chatted on the front step next to a pile of dried old whale bones that had been collected from the beaches nearby. He was used to island life, having moved to Coll from the Isle of Man, and the sound of Hebridean corncrakes had filled his ears for more than six summers.

'People come here for the peace and quiet, the scenery and the nature – and to hear and see corncrakes,' he said. 'It's a very special bird, an iconic farmland species, but those who haven't spotted one before often don't have a sense of scale. They imagine they are larger, the size of pheasants, and until they see a head poking out of the low grass they can take a while to get their eye in.'

The sympathetic management of the area for corncrakes was helping plenty of other birdlife, he told me, and a lack of feral cats, brown rats, hedgehogs, stoats and other ground predators of eggs and young reduced losses. At one stage only six corncrakes made their home on the reserve. Now, anything between fifty and seventy could be heard calling, according to the results of exhausting night-time surveys carried out every June.

Since turning up here I had heard several corncrakes and enjoyed a brief sighting from a few yards away. I didn't expect it could get any better. But I was mistaken. Ben pointed out a group of three people wading through an overgrown area and making their way up to a Land Rover parked close by.

'You're in luck,' he said. 'Looks like they've caught a couple.'

Professor Rhys Green is a leading research scientist for the RSPB and a national expert on corncrakes. He has monitored populations extensively and used geolocator tags to track their migration, discovering that Coll's corncrakes spend the winter in West Africa and forest clearings in the

Congo. Back on the island for a spot of survey work with two assistants, he just happened to have been busy trapping while I was meandering around the reserve, and as he neared I could see he was clutching two cloth drawstring bags.

Introducing myself, I joined the group at their vehicle as they prepared to examine the catch, and watched as Rhys put his hand in a bag and, like a magician, pulled out a corncrake. I wanted to applaud. He held it up for me to see. It was quite beautiful close up, with a brushstroke of light grey above large, watchful eyes, and warm hues in the plumage. It was also remarkably well behaved in the hand, and I was able to run a finger over its soft dark crown. I couldn't help myself. I mean, it's all very well hearing and seeing one, but how many people actually get to touch a corncrake?

The weight and pale facial markings indicated it was a female, about a week off laying the first of her two summer broods, and pointed wing-feather tips signified that she was a year old. This, I was surprised to learn, is the average life expectancy of a corncrake. It is often assumed that birds calling year after year from the same spot are the same individuals. In fact, they are most likely to be different generations of offspring, each fortunate to have made it back from Africa and unlikely to do so the following spring.

'It is a very unusual bird in that it puts all its effort into breeding, and virtually none into surviving,' Rhys said. 'Simply keeping populations stable means you have to work hard with farmers to ensure the right conditions are maintained and corncrakes can raise enough young. You can't just create a bit of habitat and walk away hoping for the best.'

Catching corncrakes by day involves flushing them between low nets that funnel into a mesh basket. According to an intriguing bit of research, a recording of an old Irish tractor engine has proven the most effective noise at scaring them into the trap – which, vehicle nationality aside, makes perfect sense given the species' history of farm machinery-related deaths.

The bird conjured up from the second cloth bag was a year-old male. He looked to be in prime condition, ready for a strenuous few months of vocal jousting in the battle to secure a mate – even if it meant stealing one from a rival male. Plenty of sexual shenanigans go on in the hay fields after dark.

Once the birds had been ringed and released, I left the reserve and returned to Arinagour. A car stopped and offered me a lift, even though I hadn't raised a thumb, and I was glad to accept. It had been a long day, and I would have fallen soundly asleep that night in the warm and comfortable bunkhouse dorm, but fellow inhabitants had other ideas. About half an hour after the lights had been switched off, a low drone of snoring began from the far end of the room, rising through the fug of damp hiking socks. Then another sleeper started up, like a natterjack toad answering, until the room was reverberating. I stuffed tissue into my ears and managed to nod off.

The following day, I explored the central moorland and western side of the island. I heard two corncrakes, saw a breeding pair of red-throated divers on one of the lochs, spotted a hen harrier flying over higher ground and was treated to a sighting of an otter diving for fish on the far side of a bay. The special species just kept on coming, and I toasted my successful trip in the hotel bar that evening.

Next morning, the ferry was delayed and it was uncertain when, or even whether, it would arrive. It was having engine trouble, the shop owner told me, and I might as well head off and listen out for updates. From whom I wasn't certain, but rather than waste several hours by the jetty I took her advice and hitched a lift back to the reserve for a quick fix of *krrek*. Ben told me there was no news to report on the ferry's arrival, and, walking back, a driver confirmed the same. 'No news' travels fast on Coll. There was no news at 11am, no news at 1pm and no news at 3pm. Finally, I learnt that a replacement boat was on its way. It arrived at 8pm, and although I was glad finally to board, I wasn't really in any hurry to leave. I knew, as I sailed back across to the mainland, that it would be a long time before I heard corncrakes calling again.

# Seven

So this was it. The lair. It was the first time I had actually ventured inside. Uncertain what I might find, I stepped carefully over thick mud churned up by livestock at the open entrance. There were two rooms, one on either side, with bare stone walls and missing partitions. Wind whistled through broken windows. The floors were covered in a slurry of soil and cattle dung, and a fireplace caked in dirt was filled with nesting material that had tumbled down the chimney. Abandoned for some time, this derelict farm dwelling now provided shelter for Galloway cows that grazed nearby in this remote Scottish valley.

Opposite the doorway a wooden staircase led up under the warped roof and I climbed it carefully because it was rotten and in danger of collapsing under my weight. Light streamed between missing slates revealing an open area of floorboards, damaged in places by fallen chimney blocks and carpeted in bird droppings, feathers, straw and dust. Clinging to a railing and unable to stand up fully in the confined space, I stepped lightly on the creaking boards, ducking under cobwebs and turning to take in the scene. I knew that this was the exact spot where it had been – curled up in a dry corner out of the wind, before being disturbed by approaching footsteps. The remains of a hare that had been left on the floor were long gone, and there were no footprints to be seen. Come the winter this might once again act as its temporary refuge. But, for now at least, jackdaws inhabited the upstairs area. I could hear them calling from the stand of pine trees outside, anxious for me to leave.

I climbed down and headed back into the open air, strolling up to the main path and pausing to take in the view. It was here, in the recent past, where I had encountered Scotland's most iconic predator, an animal that helped inspire my current quest to find our rarest species: the Scottish wildcat. And the atmosphere of this place, deep in the Cairngorms, and the memory of such a special sighting had remained with me ever since. I had been determined to return. Wildcats still roamed this area, hiding out in the forests that lined the ridge and hunting in the tussocky fields for rabbits and rodents. Not that I expected to get lucky a second time. It hardly mattered. I just wanted to walk once more in their territory, to see the upland valley free from snow, to view the isolated retreat and add a few more pieces to the jigsaw of my recollections.

I had come here three years earlier in search of this scarce and secretive species. Wildcats had always captured my imagination and I had decided that it was high time I treated myself and tried to find one, with my youngest daughter unexpectedly agreeing to accompany me, even though I warned her that the odds of a sighting were stacked against us. The wildcat is probably the most challenging of all our mammals to track down. Rare, silent, solitary, wary and unpredictable, it ranks on the ease-of-detection scale somewhere between 'invisible' and 'mythical'. Obviously expert help was required, and after flying to Inverness my daughter and I had met up with ecologist Adrian Davis, who had coordinated a three-year survey of their distribution. While he knew the promising areas to search, which were thinly spread across the northern half of Scotland, he was also well aware how seldom people came across them. Nevertheless, he was willing to act as our expert guide.

In simple terms the Scottish wildcat resembles a large and rugged version of a typical grey-brown tabby. But one look at a photograph and you can see that they wouldn't take kindly to having their chin tickled. They have the spirit of wilderness running through their veins. It is said that they can never be tamed in captivity, and this reputation for ferocity and independence has earned wildcats pride of place on clan emblems, such as that of the Macphersons, whose ancient 'don't-mess-with-us' motto reads: 'Touch not the cat bot [without] a glove'. Scaredy-cats be warned.

Unapproachable they may be, even savage when cornered, but wildcats have plenty to fear. These wild inhabitants of wild places have been persecuted mercilessly down the centuries and killed in unimaginable numbers. Back in the sixteenth century, at a time of serious food shortages, the species was included on a lengthy official list of animals deemed a pest, with a bounty placed on its head – its actual head. Under the Tudor Vermin Act, parishes paid upwards of a penny for every wildcat head brought in to the churchwarden – a tidy sum during this period of extreme rural poverty – and more than three hundred payments were made in Devon alone, which goes to show how widespread and numerous they once were.

The situation worsened during the Victorian era as field sports grew in popularity. Gamekeepers armed with guns and snares were employed to rid shooting estates of anything that competed for grouse and other quarry, proving ruthlessly effective. Records for one estate show that 198 wildcats were killed over just three years, and this unremitting onslaught, coupled with a loss of suitable woodland habitat, resulted in their extinction in England and Wales by the end of the 1800s and depleted numbers in Scotland. It looked as though the long war waged against this predator would finally be won, and that it would follow the lynx, wolf and bear into our history books. Yet, paradoxically, it was war that may have saved them from extermination altogether. Gamekeepers well able to handle a weapon were packed off to fight in the trenches during the First World War, and far fewer of those that returned were subsequently employed. The wildcat eventually began to spread back from the remotest recesses of Scotland, and by 1988 it was afforded legal protection.

They remain rare, and the most significant threat to their future is now believed to be domestic moggies. Despite physical similarities, they are evolutionarily distinct. The native Eurasian wildcat has been roaming these islands since at least the last ice age, while our pampered pets, with origins in the Middle East, arrived about two thousand years ago. Yet they are closely related enough to mate and, with so few wildcats around, this interbreeding threatens to hybridise our precious remaining pure-breds, numbering perhaps as few as four hundred, out of existence. As a result, schemes have been introduced to neuter domestic and feral cats in areas where they come into close contact with their endangered relatives, in the hope that this will prevent them from diluting the shrinking gene pool still further.

Even with all the cross-couplings, the Scottish wildcat has just about managed to endure as a discrete species, with visual characteristics – such as thick fur and a banded, blunt-ended tail – that enable it to be told apart from hybrids and feral cats. But it isn't always easy to identify reliably, especially given that most sightings are fleeting and

Scottish wildcat tracks in the snow, Cairngorms

The author holding an egg of the extinct great auk, Oxford Museum of Natural History

Rare common skate catch ready to be released, Sound of Mull, Scotland

A sperm whale pays an unexpected visit off Oban, Scotland

Enjoying a close-up view of a great crested newt,
Cambridgeshire

Getting to grips with a smooth snake, Dorset

Duke of Burgundy butterfly poses for the camera, Sussex

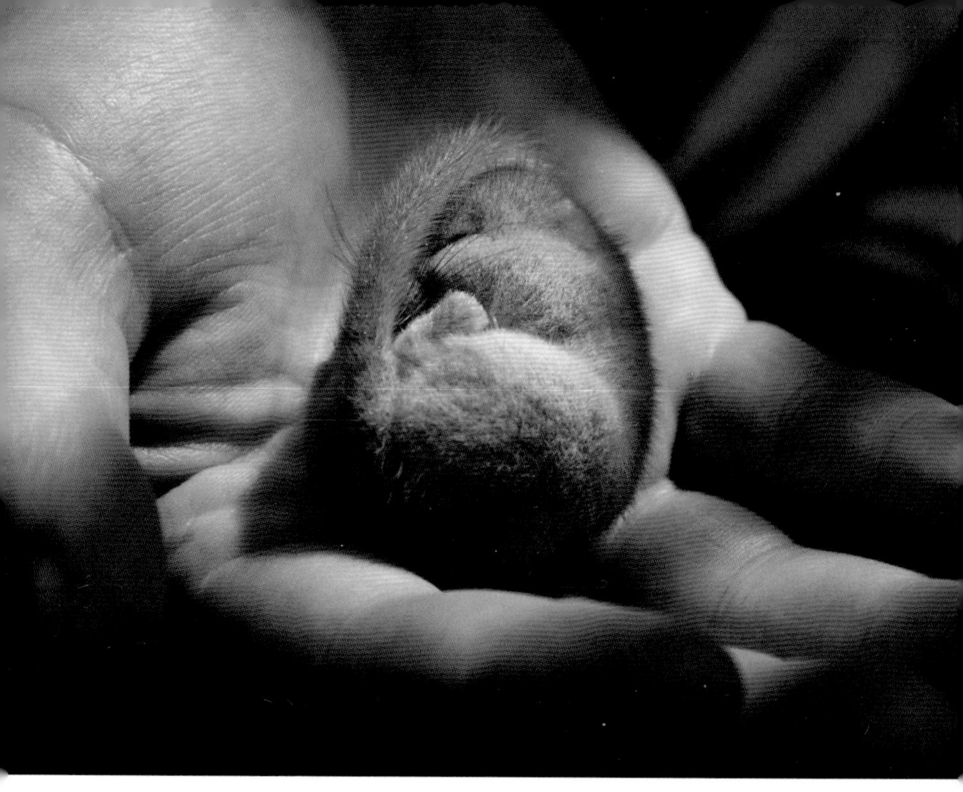

A snoozing dormouse on the Isle of Wight

A water vole swims between grassy banks at Lakenheath RSPB reserve, Suffolk

'Eel man Pete' with a catch, Norfolk

Natterjack toad at the edge of a shallow pool, Cumbria

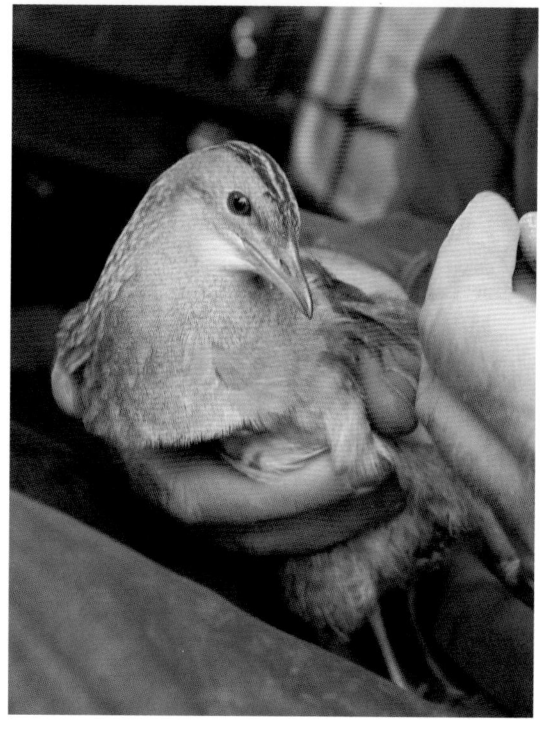

Elusive corncrake ready for ringing, Isle of Coll, Hebrides

Slavonian grebe nears the shore of Loch Ruthven, Scotland

Resident Scottish wildcat at Highland Wildlife Park, Kingussie

A pine marten pays a brief visit to a hide, Cairngorms

Impressive capercaillie in full display, Abernethy Forest, Scotland

Secretive Bechstein's bat bearing an identity bracelet, Dorset

Streaked bombardier beetle tracked down at a derelict docklands site in east London

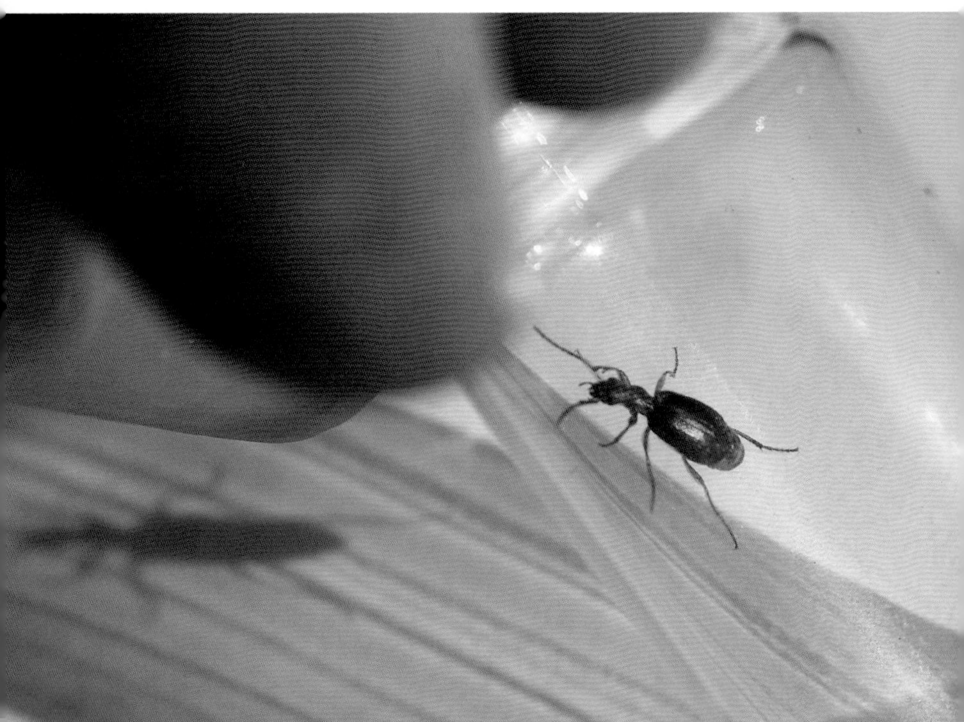

Huge basking shark passes the boat, with tail and dorsal fin visible, Isle of Man

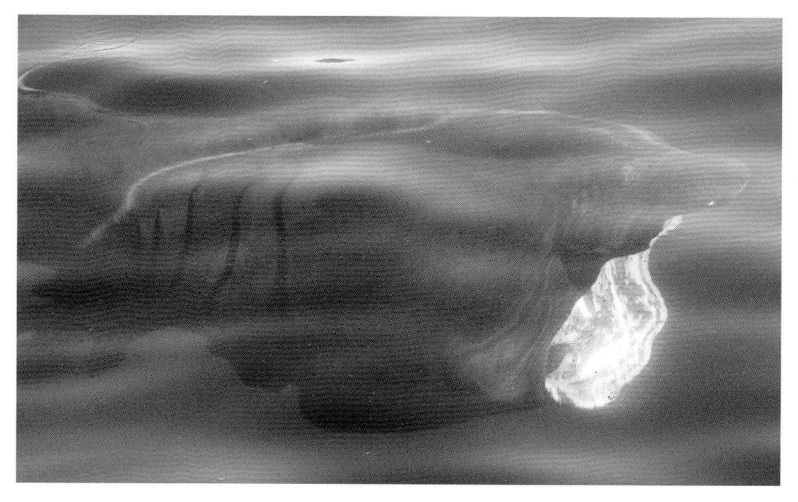

Close-up of a feeding basking shark as it passes the boat

Swabbing basking shark dorsal fin for DNA samples, Isle of Man

Pool frog at a secret location in East Anglia

Norfolk hawker dragonflies rest amid reeds, Norfolk

Fen raft spider feels for water surface vibrations of potential prey, Suffolk

Queen shrill carder bee feeds on nectar, south Wales

A black rat emerges after dark, Shiant Isles, Hebrides

Wart-biter bush cricket chirping in sunshine, Kent

Spiny seahorse hiding amid eelgrass, Studland Bay

Rare vendace fish at Derwent Water, Lake District

Black-winged stilt turning eggs on its nest, Sussex

Silurian moth at dawn, Wales

in poor light. Searching an area where genuine wildcats have been spotted narrows the chances in your favour, and I was pinning my hopes on Adrian's local knowledge.

We had stayed at his guest house in Birnam, Perthshire, and braved the freezing February weather to visit a number of areas at dawn and dusk over two days without any luck, before finally heading to a prime location near Tomintoul in the Cairngorms. Snow had begun falling by the time we reached our accommodation and we woke the following morning to a pristine landscape marked with animal tracks. The open fields resembled white pages scribbled with nocturnal comings and goings. Red deer trails could be seen in the deep drifts, the surface scuffed by their hooves, while pockmark prints of rabbits ran one way and another beside buried hedgerows.

It was still, clear and cold. The only sound was the soft crunch of our footsteps as we followed a winding uphill path away from our lodgings, scanning the ground ahead of us. The conditions were perfect. Everything that had moved in the night had left its signature in the snow: a hare breaking cover, a roe deer leaping a gate, and . . . a single set of prints brought us to a halt at the edge of a copse. It wound beneath the scattered birch trees, tracing the line of an old ditch before passing through a wide-mesh sheep fence and out onto open terrain.

Bending down, we examined the marks closely. Circular indentations with four front toes and a central pad were clearly visible where the feet had found a firm footing. Adrian looked up and nodded. 'This could be what we're looking for,' he said with a smile. 'Scottish wildcat.'

It was obvious that the animal knew where it was going. The tracks took a determined route, tracing the contours of the valley side and remaining close to boundary walls and sheep fencing, where tussocks of grass provided a little cover.

After half a mile we came across the ruined building overlooking a field and stream below. Adrian stopped and surveyed the entrance with his binoculars. 'The tracks lead straight in through the front,' he said under his breath.

Signalling for us to remain where we were, with a view of the rear of the building, he approached, skirting the side before disappearing from sight. The resident jackdaws circled above, calling, alerting anything slumbering inside to his presence.

We knelt in the snow and remained as still as we could, waiting to see what, if anything, would happen next. Then, suddenly, there it was: a large cat jumped from a broken rear window onto the ground and leapt up the sloping shattered remains of a wall into a hole between some slates and vanished.

'Did you see it? Did you see it?' I turned to my daughter, who was nodding with a big grin. My heart was pounding. 'It must still be hiding somewhere in there. I guess it didn't want to break cover and run out into the open.'

From where we were sitting at a distance it had been impossible to discern any markings because the fast-moving cat looked so dark against the snow.

Adrian returned and I rambled on excitedly about our sighting. He had heard the noise of it moving on the floorboards above him and had taken photos of the remains of prey found inside the building: the leg of a hare and feathers from a bird of some kind. Any cat that can kill a hare is certainly a serious predator.

From that tantalising glimpse I could never say for certain that it was a wildcat, although its size, local records and the out-of-the-way location certainly strengthened the case in favour. But the visit had one more surprise.

At dusk, we returned to the valley and sat downwind at some distance from the ruin, watching in the hope that the cat would eventually leave its retreat and set off for an evening's hunting. The temperature dropped quickly with the sinking sun and it was hard to bear more than an hour out in the open. With gloved hands stuffed deep in pockets, knees drawn up to our chests and teeth chattering, we patiently scanned the area in the dying light for movement. Nothing stirred. The wildcat was obviously staying put until nightfall. But I was lucky. Over the stream on the other side of the valley, far from the dilapidated building, a low shape shifting slightly from right to left caught my eye – little more than a dot in the white expanse. Through my binoculars I could make out that it was a cat crossing the frozen ground towards a pine tree. Again, it was impossible to observe any detail, but the thing that was so striking was how solidly built it was, and the way it shouldered its way through the deep snow, appearing totally indifferent to the conditions. I had never seen a cat behave in this way before. It must have climbed the tree, because it didn't continue past and, although we waited for a while, it failed to reappear and we were forced to call it a day.

There is an expression sometimes used regarding tricky birdwatching identification in the field: 'You'll know it when you see it.' This was a Scottish wildcat. I knew it. And I can still shut my eyes and see it, and know that it was.

So now I was back in Scotland hoping for a new encounter, and after revisiting the upland valley in the Cairngorms I drove to the Highland Wildlife Park at Kingussie for a close-up view of Scottish wildcats kept as part of a breeding programme.

I met up with keeper Rachel Williams, who told me that the resident pair were sleeping off a busy night of mating. Set among a stand of conifer trees, the enclosure was divided into sections connected by overhead walkways, and filled with boulders and branches. On one side the female was curled up just out of view, while the male, Hamish, a large brute of a cat, was dozing on a walkway. Both woke as soon as Rachel produced a bucket of raw meat. Quail, rat and rabbit were on the menu, and she entered the enclosure and placed portions on various branches.

Up and about, the ten-year-old male was an impressive beast – solid and broad-headed, with a mean stare and, as Rachel informed me, impeccable wildcat DNA.

'There is something special about him,' she said. 'He has a grumpy demeanour and always looks like he has had a really bad day.'

Despite being captive born and experiencing daily human contact over the years, he kept his distance from the keeper, and I asked her whether they were really that wild by nature.

'From the minute kittens are born, they're fearsome. They will hiss and spit, and they seem instinctively to know that you're not a part of their life,' she said. 'And the adults are unpredictable and can lash out, but generally they're not interested in us. If they could choose they would be nowhere near us.'

The wildlife park's policy is to ensure that the cats behave naturally and do not become tame, in case individuals should be reintroduced into the wild in the future. 'Captive breeding is very important,' Rachel added, 'because if the wild population crashes even further, cats like these could be the back-up plan.'

I had been told of a farm half an hour away in the Cairngorms, where a female Scottish wildcat had regularly been seen over recent years, and it meant driving through one of my favourite places in Britain, Abernethy Forest. Here the road weaves between towering Scots pines and skirts the banks of Loch Garten, looking out over the water to mountains beyond. It was a beautiful day and the setting in late May was idyllic. I took time out for a walk along the forest trails, spotting red squirrels and crested tits along the way, and sat where the sparkling loch kissed the shore, gazing out at the view of this magnificent landscape, one of the last remnants of the ancient Caledonian wildwood that once covered Britain. After popping into the RSPB viewing point to get a quick look at the resident pair of nesting ospreys, I drove on, eventually branching off on a rough track that led to a farmhouse hidden away in a woodland clearing. Here I met up with Hannah, a nurse, who, along with her children and farmer husband, is fortunate to have enjoyed some of the best wildcat views in Britain – from her dining table.

'It was early summer and I was sitting here at the table and I looked out and noticed it and realised what it was,' she told me, pointing from where we sat in the conservatory across the field at the rear of her house. 'I thought: wow – a wildcat! But I assumed I'd never see it again. Then it turned up a second time, and as time went on I began to see it more and more regularly, and the kids would spot it and shout out "Mum, there's the wildcat!" It became such a pattern that I could almost predict the next time we would see it.'

The wildcat emerged to hunt for mice and voles in the field at pretty much any time of day, she said, then disappeared into the surrounding forest – on occasion chased off by roe deer. Once, under cover of darkness, it even approached the house to steal a duck carcass just a few feet from where Hannah was sitting inside.

'We've been incredibly lucky to see it so often,' she said. 'And it seems comfortable with the fact we're here.'

She showed me video clips she had taken, which revealed its bulk and impressive markings, and photos of a visiting BBC wildlife film crew who got rare footage of the cat hunting. And I thumbed through diaries in which she had recorded sightings: '7am, hunting in field', '4.30pm, cat in field for over two hours', 'carrying prey', 'mobbed by crows', 'seems well and to have survived prolonged spell of frost', 'saw cat's eyes with torch while out feeding calf', 'lots of prints in the snow'.

Unfortunately there was no snow to help me this time, so I would have to rely on patience and good luck. Hannah kindly suggested I wait in the woods opposite, and after gathering a few things from my car I set off, headbutted on my way by one of their goats.

I took a path through the birch trees, determined to be as stealthy as possible, treading quietly on the soft moss, moving slowly and pausing to look around me after every few steps, like some commando on jungle patrol. Of course, the wildcat was probably watching me pass by, with that aloof and indifferent look cats have, wondering why a human was making such a big deal about walking through a wood.

There seemed to be clues everywhere: a hideaway under a rock where the ground had been pressed smooth; a half-eaten wild bird's egg; some kind of scat with fur in it; claw marks in the bark of a tree; gaps where fencing mesh had been pushed apart. All very promising.

I found a good spot with views of the field and the woods behind, and waited. A woodcock flew overhead, and I heard deer barking in the pine forest and black grouse further off. I was still and silent, at one with my surroundings, blending perfectly into . . .

*Ring-ring!* My mobile burst into life.

'Hello?' I whispered.

'Hello is that Charlie? It's the printers here. We seem to be missing page six.'

'What?'

'All the others have been sent okay, but six hasn't appeared for some reason.'

'I'm not really near my computer right now,' I said, shaking my head in disbelief. In fact I don't think I had ever felt further away. 'I'm on holiday, can you try the other editorial numbers, thanks.'

'Sure, sorry to trouble you.'

Well, that blew my cover, I sighed, switching the phone onto silent and putting it back in my pocket. I explored a little further up the path before returning to my vantage point to resume my watch. One hour passed. Then a second. Then a third, by which time it was growing dark and I started feeling a little uneasy standing alone in a wood. My pesky imagination began playing games, as it had when I was out looking for natterjack toads, this time turning me from intrepid wildlife tracker into a victim in waiting. I tried to reassure myself that a deserted area of trees on a secluded hillside in the Cairngorms was unlikely to be a crime hotspot. Out here the only marauding attackers intent on drawing blood were midges and mosquitoes. But sometimes you can't stop worrying about your safety, even if it seems as irrational as a fear of Killer Sudoku. After all, isn't forestry land just the kind of place where people come equipped for a chainsaw massacre or axe murder . . . ?

I headed back to my car and slept in it for a few hours, returning at 4am. The rising sun sent paranoid terrors scuttling back into the dark corners of my mind, and I was filled with renewed optimism about my chances of seeing a wildcat. Only, it was not to be. The magical time at dawn came and went without a glimpse, and there is only so long one can stand looking at a field and forest paths, willing an animal to appear out of thin air. Perhaps it had detected me, or was keeping clear of the new stock in the field, or was in another part of its large territory. Whatever the reason, it didn't show, and by mid-morning I abandoned my vigil.

Fortunately, in this part of Scotland, if you don't find the number-one rare carnivore, you can always try for the second scarcest: the pine marten. These bright-eyed, bushy tailed, agile arboreal hunters, shaped like supersized stoats and capable of chasing down squirrels, have suffered much the same fate as wildcats. Persecution and the loss and fragmentation of woodland over the centuries pushed this once-widespread predator to the edge of extinction in Britain. It was hunted for sport and highly prized for its dense chocolate-brown pelts – a staggering 250 of which were used to line a gown worn by Henry VI – and was shot and trapped by gamekeepers. By the early 1900s, pine martens were largely confined to north-west Scotland and were even rarer than wildcats. However, they have begun to recover and are spreading back into old haunts north of the border, where they number around 3,500. There is also evidence that a few may be living in woodlands in northern England and Wales: a roadkill carcass turns up every few years, as well as the occasional dropping, which apparently has a telltale fruity bouquet, should any wine buffs wish to join the search.

Pine martens are as elusive as wildcats, but have one weakness: a sweet tooth, which means a few can be coaxed into the open at established viewpoints. Besides a typical meat-eater's appetite for rodents, birds, carrion and eggs, they are also fond of berries, so a scattering of sultanas and spoonful of jam can prove the recipe for success when it comes to trying to spot one. Consistent sightings of an animal with such star quality offer a valuable attraction in areas that depend on ecotourism for income. As such they have joined a cast of charismatic species helping make conservation pay for landowners – a win-win-win situation for wildlife, nature enthusiasts and local communities.

I had booked a place on an evening trip to one of the most reliable viewing hides in Scotland, situated on the vast Rothiemurchus estate in the Cairngorms. Our group of eight met close to Loch an Eilein, and Speyside Wildlife guide John Picton walked us to the site in the rain, instructing everyone to keep talking to a whisper.

Unlike typical bird hides, the estate's impressive wooden construction was built to be soundproof – to allow for close encounters – with a carpeted floor, curtained areas and thick glass windows on three sides that looked out across a forest clearing. The ground around the cabin was peppered with peanuts and sultanas, and treats had also been placed on a raised wooden platform.

Pine martens tend to keep to themselves, meeting up only to mate or in territory disputes with rivals, and John told us that the current regulars were a resident male and female, along with a solitary male imposter from a neighbouring range. Individuals could be recognised by the shape of their creamy yellow 'bibs', and a wall chart showed the throat markings of various pine martens recorded in the area over the years.

'They are great to see – striking looking, full of character and quite bold, but they are not creatures of habit,' he told me. 'You can pretty much set your watch by the badgers that visit this hide, but the pine martens are very different. The male will turn up every night for a week, and then not at all for a time, perhaps because they have such large territories, while the smaller female is more reliable, though less so when she has young.'

We spread out and took our seats by the windows, while John kept his eye on a screen showing live footage from four infrared cameras trained on well-used animal tracks leading towards the hide. As the sun began to sink behind the treetops, it felt as if theatre lights were being dimmed, and we waited for the performers to take the stage.

So how long can an excited audience remain quiet when nothing is happening? About ten minutes. Then the rustling, throat clearing and murmuring begins. The volume rose, checked itself, then rose again – and I'm as bad a fidget as anyone else. We spotted a herd of red deer making their way light-footedly between the birch trees, and a lapwing flew over, paddling the air with rounded beach-bat wing tips. Then, just as I was about to point out a woodpecker working its way along a trunk opposite, two words uttered by a man to my left silenced me and everyone else in an instant.

'Pine marten.'

I don't think anyone had expected one to appear so soon. It ran along a track by his window, then around the side of the hide, until it arrived right in front of me and began tucking into the snacks on the wet grass less than five feet from where I sat, obviously totally unaware of our presence.

'That's the resident male,' John whispered as people gathered at my end of the window to get a decent look.

There is nothing like your first sighting of a new species in the wild, and this was an incredible view. He was a little damp, but still extremely handsome, and larger than I had imagined, with the kind of soft-looking fur you would willingly risk losing a finger to stroke. Although his face was pointed, the expression was more inquisitive than mean, and prominent rounded ears added to his appeal. He ate fast, darting glances around him and turning to reveal a straw-coloured throat, which contrasted smartly with his brown coat. Something was making him nervous, though. A scent of fox, perhaps, or badgers. Whatever it was, he didn't hang around, and after a minute or two he bounded off across the grass in front of us, his bushy tail disappearing into the bracken.

The sighting may have been brief, but what a privilege to have witnessed this rare animal of the ancient woods, like a visitation from our medieval past.

There was a heightened sense of anticipation among the group. It seemed anything could happen at any moment, and everyone diligently scanned the hide surroundings in silence. Ten minutes passed, he failed to return, and the whispering resumed. However, the badgers arrived 'right on cue': a dominant female and broader-muzzled male trotted up the path and began hoovering up peanuts beneath the main window. They were fascinating to observe at such close quarters, and we waited for them to finish and move on before we could leave, returning to our cars delighted with the evening's entertainment. As wildlife shows go, this is one you hope will run and run. And it could be coming to a hide near you if pine-marten reintroduction plans for other sites across Britain get the go-ahead.

The Scottish wildcat, on the other hand, was suffering from stage fright. I made two further return visits to the hillside site and endured long waits in the woods, but it continued to sulk in the wings and refused to make its presence known. However, I did get a surprise on the drive there one morning through Abernethy Forest. As I rounded a bend, something caught my eye and I slammed on the brakes and pulled in quickly, turning off the engine and grabbing my binoculars. I hardly needed them. A few metres away, circling slowly on the ground beneath the trees with tail fanned turkey-style and head held high, was one of Scotland's most famous birds in full display: a male capercaillie. And this giant glossy black member of the grouse family, pumped up with testosterone, was so preoccupied with strutting his stuff that he scarcely noticed me getting out of my car to draw nearer.

I wasn't alone. I noticed a man crouching beneath one of the pines, watching, and I joined him. It turned out he was a keen ornithologist from France, who had been as surprised as me to come across such an astonishing and seldom-seen wildlife spectacle.

'C'est incroyable,' I said in my best French accent.

'Non,' he replied, looking puzzled. 'C'est un capercaillie.'

We sat side by side, leaning back against the roadside bank, staring in amazement at one of Britain's most remarkable birds doing what it does best: showing off. And it did so in an unhurried fashion, taking a few steps between the heather and bilberry, then spreading its tail and stretching its head upwards to reveal a ragged black beard of feathers, while uttering a few guttural sounds and clicking its beak like pruning shears. The soft light that filtered through the tops of the tall pines hung like a mist between the lichen-mottled trunks and gave the scene an enchanted quality. It took a few more steps, pacing the display ground, puffing out its solid body and demonstrating its dominance should any females or rival males be close by. All the while it observed us from under red eyebrow wattles, looking far more likely to attack than flee. This brute of a bird that eats pine needles for breakfast was not to be messed with.

Tough it may be, but sadly the capercaillie is facing extinction in Britain – for the second time. Over-hunting and extensive tree felling spelt its demise by the late 1700s, and while the birds reintroduced soon after multiplied to reach a population in the tens of thousands, loss of habitat and other factors mean the species is now in rapid retreat and down to just twelve hundred. To lose them again would be, quite simply, a tragedy.

My parked car had obviously drawn the attention of others passing along the road because birdwatchers began to gather on the verge to share in our good fortune. Out came the telescopes, the cameras and the mobile phones, and after a quarter of an hour at least a dozen spectators had assembled to admire the bird. I was glad so many got the chance to see it before it gradually moved away and out of view. Some experiences should be enjoyed by as many people as possible, and nothing makes a better case for conservation than encounters with animals themselves – especially a threatened species as majestic as the capercaillie.

Uncommon as it was, the fantastic capercaillie was not on my target list of rare birds. I was hoping to find an even scarcer breeding species, and the following day I said goodbye to the forest and Cairngorm mountains and headed west to Loch Ruthven, a freshwater lake set among birch woods and lush pasture near Loch Ness.

We tend to think of birds as born to fly, at home in the vast blue firmament. But for many species, from partridges, grouse and guillemots to woodpeckers, divers and rails, taking to the air is either a last resort to avoid predators or a necessity to get from A to B, rather than a way of life. You get the feeling that corncrakes, for example, able to fly here from Africa, would grumble the whole way on migration 'are we there yet?', finally arriving with a sigh of relief that they can avoid getting airborne again for a few months. And among those groups content to leave the heavens above to happy-flappy birds are grebes. Members of this family are built for a life in water rather than off the ground. They are even pretty ungainly on land, with feet set far back on their bodies to aid propulsion when swimming and diving. Covered in dense waterproof plumage, they have thin wings, elongated necks and sharp pointy bills to outmanoeuvre small fish and invertebrate prey, and are so adapted for aquatic existence that they build floating nests of weed and plunge underwater when alarmed. High-flying masters of the skies they are not. Stick a grebe at the top of a tree and it would probably suffer from vertigo.

Loch Ruthven is the national stronghold for arguably the most attractive of our five species of grebe: the Slavonian grebe. In winter, when these moorhen-sized birds can be found around the coast, they are black and white and fairly plain looking. In spring, however, they moult into breeding finery and are completely transformed. The virtually tailless body turns a combination of rusty brown and charcoal grey, and a mane of glossy dark feathers enlarges the appearance of the head. To top it all, sprays of golden plumes extend back from cherry-red eyes in two striking tufts, giving rise to its alternative, and more sensible, American name: horned grebe.

Back in 1796, a specimen killed near Truro in Cornwall and about to be plucked by a local fisherman was brought to the naturalist George Montagu, who was in the process of compiling a comprehensive ornithological catalogue of British species. In describing the discovery, he noted that it was also understood to reside in North America and 'Sclavonia' in Europe, and he labelled it the Sclavonian grebe. Whether this referred to Baltic Scalovia, or the Slavonia region of eastern Croatia, is unclear – and let's not throw Slovenia and Slovakia into the mix. Either way, eventually it became known as the Slavonian grebe, and this strange name has stuck.

A little over a century later, these northerly breeders were found nesting in Britain, and it seemed that well-vegetated lochs were to their liking. Numbers crept up, until by the 1980s Scotland could boast over eighty pairs. It was still rare, but heading in the right direction. Then its fortunes began to take a turn for the worse. No one is exactly sure why. Every theory that makes sense on one lake, such as the presence of trout and pike, human disturbance or changing water heights and nutrient levels, seems to be disproven on another, and counts have even fluctuated on lochs where conditions remain stable. Whatever the reason, the UK population nosedived to a current level hovering at just over twenty pairs. You could fit all of our breeding Slavonian grebes into a supermarket bag for life.

I'm not surprised that several pairs make the effort to fly to Loch Ruthven for the summer. The reserve is a lovely, tranquil spot tucked between lightly wooded hills, moorland and green fields. Worthy of its Scrabble assortment of conservation designations – SSSI, SPA, RAMSAR, SAC – the scenic setting attracts ospreys, black-throated divers and a steady trickle of birdwatchers, who come to admire the grebes found near the RSPB hide a short stroll from the car park. For some it is worth a day out, while for others the reliable site offers the chance of adding a quick-fix 'slav' to their tally of UK sightings before heading elsewhere – the equivalent of ornithological speed dating. It would be wrong, though, to judge one approach as any better than the other. Even among fanatical, species-stockpiling 'twitchers', the hobby is as much about collecting experiences and memories as amassing vast numbers of 'ticks' for their lists. Ultimately, in the simple act of noticing birds at whatever speed suits, everyone is sharing in an appreciation of the variety of life.

Inside the RSPB hide I took a seat and quietly opened a hatch to survey the loch. A man and woman seated at the far end told me they hadn't seen one yet – the species they were alluding to was taken as read. It was quite breezy and I could make out a few tufted ducks riding the choppy water in the centre of the lake, but nothing else. Then I spotted a small grebe paddling about among the sedge beds just below us.

'There!' I said, pointing triumphantly.

The couple scrutinised the bird carefully. 'That's a dabchick,' the man replied.

Dabchick? He was right. I had jumped in too soon without checking it over properly. In my defence, I don't come across many grebes in my day-to-day life, and hadn't expected another kind on the lake to confuse matters. Still, the dabchick, a widespread species also known as a little grebe, lacks the distinguishing feature of its close relative the Slavonian grebe: golden head tufts. Time to keep quiet and keep looking.

A short while later I saw a similar bird quite a way out, diving near the ducks. It certainly looked promising, with noticeable yellow on the sides of its head, but I kept my mouth shut, waiting for it to surface again.

Up it popped. 'There's one,' the woman said, 'by the tufted ducks.'

'Well spotted,' I said.

'And another to the left,' the man added.

Now they were showing off.

The two grebes floated low in the water, with neck and head held up like periscopes, and dived frequently, always reappearing further away than one expected. They were clearly a pair, remaining close to one another, and bit by bit they drifted towards us until they were just a short distance from where we sat. It was an excellent view. I could clearly see the glowing red eyes and flames of yellow feathers streaming away like comets blazing across their dark heads.

One of the grebes dived and came up with a piece of weed in its bill. Grebes are known for their elaborate courtship displays, which include weed carrying, but the mate showed no interest and the green strand was soon abandoned. Several more birdwatchers joined us in the hide, and we were treated to a second pair of Slavonian grebes that appeared from around a bend, seeking out the sheltered water, followed by a single bird.

I had to leave, and walking to my car I counted another, bringing the morning's total to six – quite a sizeable proportion of the British breeding population. Unfortunately it was time to head home. Back to work. And there was also some unfinished bat business to attend to.

# Eight

I have no great interest in cars. I understand that they come in different colours and that there are various companies that make them, but that's about it. And, living in a remote village on Dartmoor, where everyone needs to get behind the wheel even to buy a pint of milk, this can be a real shortcoming. Recognising neighbours means remembering what they drive, as you are far more likely to pass them in their car than on foot. So, given that I have a brain as absorbent as post-war toilet paper when it comes to vehicle makes and models, I have adopted a policy of raising a hand to greet every car I pass within a five mile radius of home. And I'm sure others do the same. Tourists driving across the moor must mistakenly imagine all the waving locals are being friendly and welcoming. We're not. We're just terrified of accidentally snubbing people we know.

What with the hand raising and smiling, I like to think that over the years I have avoided making too many enemies, which is especially important being someone of such standing within the community – well, based on my surname. I am, after all, the 'village Elder'. Not that anyone comes seeking my guidance: 'My horse is lame, what shall I do?' 'My wife is barren, what remedy do you suggest?' No, the only questions I get asked tend to be about birds, and, curiously, everyone wants to have spotted something rare, regardless of whether they are interested in the subject or not.

Given that how you behave is subject to constant scrutiny in a small community, you have to try your best to act like a perfectly normal sort of person, which takes quite a bit of effort over the long term, and I began to wonder what conclusions neighbours might draw from my own behaviour over the preceding months, considering that I could regularly be seen throwing a duvet and spare clothes into my car and disappearing at any hour. I'm sure they had begun to question either my sanity or my marriage, and the explanation that I was going in search of a scarce newt, toad or bird might not have helped convince them that all was well on either count.

That said, I had to leave everyone guessing, because the summer meant plenty more time on the road. The Bechstein's bat, which had successfully given me the slip so far, was back in my sights once again, top of the priority list. I found out that Daniel Whitby, who I had met in Sussex, was joining Colin Morris at bat-box-filled Brackett's Coppice in Dorset to carry out a night survey of the wood. The opportunity to accompany both experts back at this prime site was too good to be missed, and I headed there directly after work.

Dan and Colin were waiting for me at the gate, along with three local bat-group members, who had turned out to help. We followed a path down between the trees then split into pairs to set up a combination of harp traps and mist nets across any open corridors in the woodland that might act as flight paths. The air was heavy with forest scents: mould, sap, rotten timber, leaf litter and, most pungent of all, wild garlic, dashing any chance of catching vampires.

Assembling by a stream, we waited until nightfall, which comes on fast under the trees. Soon the head torches were on, the midges were out, and bats could be seen circling beneath the canopy. Time to begin checking the traps, with any finds to be brought to base camp at the stream.

Damp foliage and thick mud proved a slippery combination underfoot, and I managed to fall over a couple of times on the steep tracks as I followed first Colin to check a net then Dan to inspect his harp traps. The warm conditions seemed promising and it wasn't long before we had a few bats bagged up ready for recording.

'Right, what's this one?' Colin tested me as he pulled out the first of the night, illuminating its tiny form with his head torch and carefully spreading its wings.

'A pipistrelle?' I answered.

'But which one?'

'Well, it has quite a dark face, so I'm going to guess a common pipistrelle.'

'Correct.'

Blimey. I felt quite chuffed I had got it right, and glad to be back among bats and bat enthusiasts once again, watching and learning.

The group then examined the other catches, and a Natterer's bat and two soprano pipistrelles were added to the list before everyone headed off for another round of checks. This time, a noisy noctule and a brown long-eared bat were recorded and sent on their way, and, after an in-depth discussion on the finer points of Leisler's bat identification, the traps were revisited, with people somehow managing to retrace their steps through the wood in the dark. I tagged along, helping as best I could and praying that the rarity I sought would eventually turn up. Bechstein's bat is not the only scarce species in Britain, but there is something very special about this secretive resident, with its distinctive appearance and enigmatic habits, which has made it such a conservation priority and significant find. Everyone was hoping we would get lucky.

Despite the late hour, none of my companions' energy seemed to flag. However, as time passed, my optimism gradually began to wane. I had put plenty of effort into trying to see a Bechstein's bat in the wild, and was facing the possibility that yet again it might not happen. Perhaps it had been decreed by the bat gods, though I tried not to dwell on what disturbing black-cloaked form they might take.

Then Dan returned from checking a trap and reached into one of the cloth bags he was clutching. 'I might have something for you,' he said.

His smile gave it away.

'Any idea what this one is?' he asked, gently lifting out a bat and holding it towards me.

Yes! There it was, furry body cupped in his grasp, bald hyena-like muzzle poking out over his thumb, long ears erect, staring right at me with small black eyes. I could have kissed it, if it didn't look like it did. At long last, a Bechstein's bat.

It was a mature female and she had a metal identification clasp around the forearm of her right wing, ring number Y8520. She was one of the Brackett's Coppice locals and very active, making squeaking sounds as she crawled between Dan's hands.

Colin checked her over, and after her details were recorded he opened his palm and she spread her broad wings and took off, flying up between a gap in the branches and disappearing into the darkness. It was a sight I will long remember, this unique creature fluttering away from our torch beams, back to a nocturnal life of mystery. And standing in this protected wood watching her go, among people who cared, filled me with hope for the species' future.

As if that wasn't thrilling enough, two more Bechstein's bats turned up a short while later – one in a mist net and one in a harp trap – and I was able to get a good look as they were examined. Both were ringed females, which didn't come as any surprise to Colin. His long-term study of the group in this coppice has found it to be a close-knit, female-only community during the summer months. No one knows where the males spend their time.

Finally another was caught, the last bat of the night, and she turned out to be Y8520 once again, seemingly on a mission to ensure I got to see her kind before the night was out. I was glad I could thank her twice.

If you came across half a dozen people wearing white coveralls and protective face masks, searching amid the rubble of a wasteland site in the Docklands area of east London, you might imagine they were a forensics team investigating a serious crime. The neglected litter-strewn location close to the Thames was just the kind of place where a body could be dumped. Yet the reason for the operation sounded even more serious than a run-of-the-mill gangland murder. The target was a group that had gone into hiding, armed with chemical weapons containing a potent mix of hydroquinone and hydrogen peroxide – a compound also used in the manufacture of improvised explosive devices. Disturbing news. Even more intriguing, the squad combing the ground and risking its safety to capture the potential bombers was a crack team of entomologists. Insect experts. Had there been some intelligence mix-up over news that the site was 'bugged'?

Perhaps a sense of scale is required. The subjects being sought were not only small enough to slip through a security checkpoint undetected, but also to go virtually unnoticed in a handful of soil. At just 7.5 millimetres long, they were tiny. And their explosive capabilities, impressive though they were relative to their size, posed no terrorist threat to the surrounding warehouses and housing estates. At best they could knock an ant or two off its feet. The investigation was, in fact, a desperate attempt to save an extremely rare and remarkable creature: the streaked bombardier beetle.

The military title given to this species is entirely justified. Like their artillery-wielding namesakes in the armed forces, bombardier beetles blast their enemies, and do so by deploying one of nature's most sophisticated defence mechanisms. When under attack, the bombardier beetle releases chemicals from two storage reservoirs in its abdomen into a thick-walled mixing chamber, where catalytic enzymes prompt a violent reaction to occur. The volatile mixture rapidly reaches boiling point, building up pressure as it vaporises, and is ejected from the rear end in a powerful spray of caustic liquids that can blind small mammals and kill invertebrate predators outright.

To understand why the group of entomologists were all covered up and out looking for these miniature combatants in the heart of London in 2011, you need to roll back the years to the 1800s, when beetle collecting was all the rage. Learned Victorian gentlemen who were keen to spend hours away from home avoiding domestic duties, but who couldn't play golf, found time-consuming coleopterology provided the perfect hobby. Beetles are not only varied, attractive and easy to catch, but there are an awful lot of them to keep the amateur enthusiast occupied. In global terms, beetles account for more species than any other animal order, with four hundred thousand described so far and plenty more out there waiting to be discovered. In Britain alone there are around four thousand to get to grips with (not quite as many as our species of flies, but then collecting flies is, well, never going to have quite the same appeal).

Beetles are immensely successful insects. Armour-plated, mobile and equipped with biting jaws and a good sense of smell, they are pretty much able to survive anywhere and eat anything, and are incredibly diverse, ranging from sacred scarabs, high-speed tigers, much-loved ladybirds and imposing stags to wood borers, weird weevils, diving beetles, dung rollers and thundering goliaths. The patterning and metallic sheen of the tough elytra – those hardened forewings that encase and protect their folded hind wings beneath – can also be exceedingly beautiful. Little wonder early collectors took such a shine to these assorted jewels, filling cabinets with row upon row of specimens. The sheer multiplicity must have driven them to distraction. The acclaimed geneticist J. B. S. Haldane was once asked by theologians what could be inferred about the mind of God from the works of His Creation. He famously quipped: 'An inordinate fondness for beetles.'

Among the many Coleoptera that creepeth upon the earth, the ground beetles are a particularly large family of typically active predators that could be considered beneficial because they hunt invertebrate pests, and they include the widespread bombardier beetles, which number more than 250 species worldwide. Britain's northerly position, however, means that we are not spoilt with an abundance of these little hotshots. In fact, nineteenth-century coleopterists identified just two resident species. The first was the common bombardier, which isn't common at all and resembles a large, orangey red ant carrying a shiny blue-green backpack. The second, the streaked bombardier beetle, is virtually identical, but has a distinctive red streak down its back. It was even rarer, and collectors managed only to record individuals in Essex in 1820, Kent in 1830 and one in East Sussex, which was found in 1928 but not unearthed until relatively recently in a museum collection. With none known to have been caught since then, the streaked bombardier beetle was presumed extinct in Britain. No one had any idea that they were still clinging on – and in the unlikeliest of places.

In June 2005, naturalist Richard Jones was surveying an area of derelict land earmarked for housing near the Thames Barrier, checking under squares of roofing felt for reptiles and small mammals, when something scuttling away across the dry, weed-ridden earth caught his eye. Most of us would have thought 'ah, a pretty little beetle', if we noticed it at all. But Richard just happens to be a fellow of the Royal Entomological Society and the Linnean Society of London, as well as past president of the British Entomological and Natural History Society. Instantly he recognised that 'pretty little beetle' as an extraordinary find. As he says himself, his 'heart soared'. It was a streaked bombardier beetle, the first in nearly ninety years, and it was not alone. Return visits turned up dozens of them, living amid the rubble on the sparsely vegetated earth bank.

The joy of the discovery was tempered with the realisation that the entire area was due to be developed and the beetles bulldozed into oblivion. Something had to be done, and fast. Rocks and soil were dug up and a mound created away from the site of the planned flats, and more than sixty streaked bombardiers were captured and moved to this fenced-off refuge – arguably Britain's smallest and most unsightly reserve. The beetles like well-drained, warm areas with thin plant cover and patches of stony, bare ground, as well as the company of plenty of other beetle species on which their larvae feed. Messy wasteland sites can provide for these requirements; however, bombardiers are also quite fussy. Keeping the conditions exactly right at this rapidly constructed sanctuary proved difficult, and the numbers caught eventually tailed off to zero.

One beetle was found nearby at Mile End Park in 2010, but that was it. Things were looking desperate for Britain's rarest beetle. Then, a year later, a population was discovered right in the path of a planned development of bars and venues catering for visitors to the London Olympics. Once again, conservationists, unable to halt the scheme, had to save and shift as many as they could to the safety of a new mound piled up at the University of East London. To do so meant donning those white coveralls and masks, because the land was contaminated with asbestos. Eighteen were caught after a painstaking fingertip search and relocated – but only one briefly sighted five months after release.

Lost and found, then lost again, it seemed as if the streaked bombardier was doomed, until the invertebrate charity Buglife chanced upon several living on an overgrown rubble heap a mile away near the offices of Newham Borough Council in 2012. The threat of development hangs over this Docklands site as well, but, as things stand, occasional finds mean this remains our only stable population of streaked bombardiers.

I took a train to London to join Buglife lead ecologist Sarah Henshall on her first site check of the summer. During the journey up I was fortunate to spot a pair of red kites swooping close to the railway embankment, and wanted to point them out to passengers sitting near me, but everyone was glued to their tablets, laptops, mobiles and Kindles and the large window screens showing the real world outside appeared to be of little interest. The best way to get people's attention might have been to drag a finger across the glass and spread my hand, as if enlarging the view on a giant iPad. From Paddington I took the Tube, then hopped on a Docklands Light Railway train, which ran beside a muddy stretch of the Thames and through a jumble of architectural ideas before arriving at Royal Albert station near London City airport. It was a long way to come on a beetle hunt, so I was a bit disappointed that I didn't get collared by some market researcher asking why I was visiting the capital.

The scenic Cairngorm mountains, Dorset heaths, open fenland and windswept Hebrides all seemed a distant memory as I gazed out from the station at dual carriageways and new buildings, the air filled with the roar of planes taking off from the runway opposite. It was an improbable place to find myself on my wildlife travels. Added to that, I didn't feel particularly optimistic about my chances of seeing a streaked bombardier, even though the weather was good and I was accompanying the expert who first uncovered them at this key site. When compiling my target list of rare animals, this particular species had always worried me: too small, too scarce, too elusive – too likely to have become extinct before I got the chance even to look for it. No one had surveyed this site for at least six months, and I was well aware that good news about new colonies had invariably been followed by bad. What with all the population ups and downs, the streaked bombardier beetle wasn't so much living on the brink in Britain as bungee jumping off the edge.

Sarah arrived and led me a short distance to an overgrown area of broken paving and gravel that lay between a main road and a slab of flat brown Thames. In front of us was a ten-foot-high bank of discarded rubble and soil that had been knitted together by plant growth and was covered with various grasses and flowering daisies and ragwort. About the size of two tennis courts, it was surrounded by a loose cordon of plastic orange netting, mostly lying on the ground where the metal stakes had toppled over, and a laminated A4 notice tied to the makeshift fencing said: 'Please keep out, this site contains rare and threatened invertebrates.'

'Right, this is where we look,' Sarah said, giving me a plastic pot in case I found anything, and telling me to check shady hiding places under stones.

She carried a pooter – a jar with an attached tube that enabled her to suck little critters into the glass container – and I began by watching her at work. With a PhD in rove beetles a millimetre or two long, she was obviously a trained eye when it came to tiny creatures, and I was amazed at how low she got to the ground and how carefully she lifted the rocks and replaced them.

'People do ask me what on earth I'm doing grubbing around on sites like this,' she laughed. 'It can look a bit unusual.'

I began my search, stepping between piping, thistles, twists of wire and chunks of asphalt, and lifted up a broken breeze block that lay amid the long grass. It was astounding how much life lurked beneath: ants, snails, spiders, woodlice, centipedes and black ground beetles that scurried away, full of energy on this hot June day. I looked under bricks and cracked paving, remains of old carpet and pallets, and wished I knew more about what I was seeing. A Devil's coach-horse beetle was about the only species I recognised for certain. But regardless of my entomological ignorance I was enjoying myself and, lost in a low-level jungle, I soon forgot where I was.

Former industrial land like this, where nature has been allowed to get a foothold among derelict buildings, dumped construction materials and abandoned rail sidings, can harbour a wealth of wildlife. The infertile and seemingly inhospitable conditions of brownfield sites prevent common plants from dominating and foster a surprising diversity of flowers and invertebrates. Conservationists recognise that these unloved mosaics of mini-meadows, dunes and heaths are increasingly worth preserving.

'This is actually an endangered habitat,' Sarah said. 'Wildlife-rich urban areas are being lost at an alarming rate, and we need to view them in a more imaginative way, rather than thinking that landscaping for new developments simply means short grass and lollipop trees.'

It was hard work looking for the beetles. I had been at it for only half an hour and my bare arms were already stung by nettles, my fingers grazed by rocks and my legs bruised from kneeling. Despite the combined membership of conservation charities in Britain totalling several million, we still regard nature spotting as a faintly eccentric activity – and rootling around in this uninviting corner of London certainly had a sense of the bizarre about it. But, if you ask me, it was far more exciting than passively gawping at a distant nesting osprey on the Loch Garten webcam, and I thought luck was on my side when I lifted a stone and exposed a promising little beetle to the glare of the sunlight.

Bingo! I shouted out and Sarah came running. Only I had been mistaken. Right colours, but in the wrong order. However, it was an important prey species, which was a sign that this mound was still in good health, bombardier-wise.

A couple of hours passed and it seemed as if my initial fears of failure might be realised, with the clock ticking on the time left before my return train journey. I also had trouble locating Sarah, who was hunting so low in the long grass that she had disappeared from view. Expert entomologists are uncommon enough as it is, so it wouldn't have looked good to have lost one on my travels. Then I saw something moving, and she emerged from the undergrowth holding up her pooter jar, calling out as she walked over: 'Got one!'

Unbelievable.

'Phew,' she said, wiping her brow, looking as pleased and relieved as I felt. 'I was worried for a while.'

We peered into the container. 'Isn't it lovely?' Sarah enthused. 'There, you can see the red streak on its back.'

How she found it I'll never know. It really was small, but a gem of a beetle, with glossy green elytra, a blood-orange body and thin antennae. Not only was this a streaked bombardier beetle, the first record of the year in Britain, but, even better, it was a female heavily laden with fertilised eggs – evidence it was not alone. And she was saving her energy rather than trying to blast us with scalding chemicals through the container glass.

After observing her for a while, we returned to the exact spot where she was found, and Sarah was kind enough to let me release the beetle. I gently tipped the container and we watched as she scuttled away over the dry soil between the grass roots, carrying the next generation beneath her.

On every train journey there is typically one insane person per carriage, so to all those passengers who sat near me on the 18.35 Paddington to Exeter service that day I would like to apologise for grinning away throughout the entire two-hour journey. If it helps, I guess I was suffering from a mild case of beetle mania.

In my pursuit of rare animals so far, I had been lucky enough to enjoy encounters with creatures great and small, from the giant sperm whale and common skate to the tiny streaked bombardier beetle. I'd seen enigmatic species such as the common eel and Bechstein's bat; attractive animals like the Slavonian grebe and Duke of Burgundy butterfly; elusive scarcities, including the smooth snake, corncrake and golden oriole; not forgetting the noisy natterjack toad, eye-catching sand lizard, impressive great crested newt, characterful pine marten and cuddly dormouse, and I had walked once more in the footsteps of a Scottish wildcat. But in terms of awe-inspiring wonder, my close encounters with the next animal would be hard to beat. And the magic started with a single word . . .

# Nine

'Shark!'

Nothing can compare to that cry. Terrifying and thrilling, the short, sharp exclamation slaps you round the face and sets your nerves on edge. And no sight stirs up fears and fascinations more than the simple geometry that accompanies it: a triangular fin breaking the surface of the sea, cutting like a blade between two worlds. It was what we had been waiting for over an hour to see, and with a squeeze on the throttle Graham turned the boat and we headed out to meet the biggest fish in our seas.

I had driven to Liverpool the day before, left my car in a long-stay car park and boarded a high-speed catamaran for the Isle of Man. It was the tail end of the annual TT motorbike races and local residents were flooding home now that it was just about safe enough to cross a road again. I'm not sure whether there had been two-for-one deals on children in Liverpool, but everyone seemed to have at least three or four youngsters with them, all hyped up with the excitement of being on a boat and racing around exploring the decks.

The island's main town was busy with bikers and racegoers, and I followed the wide promenade from the port around the east coast bay, past funfair rides and fast food stands, and stopped off at a shop to buy a few unhealthy snacks for the next day, including a pack of biscuits that contained the peculiar ingredient 'exhausted vanilla beans'. I knew how they felt. I had been on the road since dawn, and by the time I found my B&B and clambered up the narrow staircase to my room I was ready to collapse. Only I couldn't switch off for worrying that the weather forecast might be wrong. It was about as perfect as it could get. A warm and windless high-pressure system was settling nicely over the island, smoothing the sea and lifting plankton-rich waters to the surface. By midday, if the conditions remained as predicted, sheltered bays should be thick with slicks of tiny organisms: shrimps no bigger than grains of rice, fish fry, copepods and microscopic larvae. This organic soup would bring in the sharks I hoped to see – in fact they eat nothing else.

What I sought was no maneater or fearsome predator, but rather the basking shark, a filter-feeding gentle giant that reaches a phenomenal size on this briny broth of animal plankton. The length of a bus, and weighing much the same, it is by far the biggest fish to be found around our coasts, and on a global scale only the whale shark is larger. So immense is the basking shark that it used to be hunted using whaling harpoons then towed to shore in order to be cut up.

It's the same old story: we caught too many. Thousands upon thousands of these slow breeders were hauled from the north-east Atlantic during the nineteenth and twentieth centuries and processed for their meat, the vast quantities of oil in their huge livers, and their rough skin, which could be used for everything from non-slip boot soles to sandpaper. Naturalist and author Gavin Maxwell, of *Ring of Bright Water* fame, was among those who targeted them, buying the Hebridean island of Soay, near Skye, after the Second World War and setting up a fishery using ex-Royal Navy gunboats to harvest basking sharks in the Minch. It was a venture that ended in frustration and failure as both catches and liver-oil prices plummeted, and elsewhere declining populations acted as a wake-up call that the once-abundant species, whose size spawned numerous tales about sea monsters, was in danger of following its own legends into the history books.

Not until the 1990s were basking sharks finally offered protection in the British Isles, and fishing for them has also been banned or restricted in other parts of the world, including American and New Zealand waters and the Mediterranean. However, despite becoming increasingly rare, they are still caught by fishing nations and their fins command large prices in Asian markets, where an appetite for shark-fin soup is believed to account for the killing of one hundred million sharks of various species every year.

In Britain, basking sharks congregate in three main hotspots in the summer: off the south-west coast, in the seas of north-west Scotland and along the western edge of the Isle of Man, the first place to introduce a ban on hunting them. They arrive from the Atlantic, where they spend the winter feeding deep in the ocean and living off their reserves, and seek out concentrations of zooplankton, in particular small shrimps, using their acute sense of smell. Opening their cavernous mouths, they plough the water, straining the goodness from it with mucous-covered gill rakers. In hot, calm weather when the sea stratifies in layers, trapping their food near the surface, the sharks can be spotted cruising along at a couple of miles per hour, seemingly 'basking' in the sunshine.

I woke early the next day to blue skies and scarcely a breeze, relieved to find the weather as forecast, and caught a bus to the harbour of Port St Mary along TT racetrack roads where every stone wall corner was cushioned with hay bales and foam pads. On the outskirts of the village I met Jackie and Graham Hall in their hillside house overlooking the bay, as they prepared for a day on the water. Two student volunteers, Natasha and Haley, had joined them in their kitchen and everyone was busy sorting equipment and logging encounters from the day before. Days, weeks, even months can go by when the conditions are lousy and basking sharks remain out of view, but when the good times roll it can be frantic for the couple – who moved to the island more than a decade earlier in the hope of enjoying a relaxing early retirement.

'It didn't quite work out that way!' Jackie laughs.

She originally volunteered to help oversee records of basking shark sightings, but extended this endeavour so that it is now pretty much a full-time job. Combining her marine biology expertise with Graham's practical background in engineering, they launched Manx Basking Shark Watch in 2005 to find out more about the sharks that ply these waters – how many there are, whether they come back year after year, where they go in the winter and whether they breed in the area – and set about producing photographic and DNA profiles of individual sharks and tagging as many as they could with satellite-tracking devices, building up an impressive bank of data that is shedding light on the behaviour of these annual visitors.

'Hopefully the information about their movements and genetics can make a real difference in their future conservation,' Jackie said. 'They are breathtaking animals to see. It doesn't matter how many years you have been studying them, when one goes under the boat and you look down at its bulk gliding beneath, you can't help but be moved by its size and beauty and the terrible shame it would be to lose them.'

Despite the demands of the work, Jackie and Graham were brimming with energy and enthusiasm, sharing jokes with the students while we loaded up the van before driving down to the harbour. Given their success the previous day, they were optimistic about our prospects, and as we boarded the six-metre research vessel *Happy Jack* and motored out of the harbour, Graham even broke into a sea shanty or two. This was serious science made fun, and I was enjoying myself already.

Negotiating the narrow channel between the southerly tip of the island and the Calf of Man, we turned north, following the western coast and keeping our eyes peeled. It was a little choppier than expected, but we caught a glimpse of a harbour porpoise and spotted a grey seal bobbing in the waves. Anything breaking the surface can be confused with a basking shark, from marine mammals to drifting tree trunks and lobster-pot buoys, and I was told to look not just for dark fins, but also for flashes of sunlight reflected off them.

What little wind there was petered out, and by the time we reached Niarbyl Bay the sea state had reached 'zero' – flat calm – and ragged trails of planktonic life were visible at the surface. I could make out tiny crustaceans, sea gooseberries and miniature moon jellyfish, like transparent buttons among myriad organisms that sparkled in the sunlight. Graham cut the engines and we drifted in silence, scanning the sea before motoring on and pausing again, motoring and pausing. Gradually we worked our way north across the bay before the cry came from Natasha: 'Shark!'

Fifty metres away a large fin broke the surface and began slowly scoring a line through the water. It is one of the most iconic sights in the natural world, an elementary symbol, branded by films and documentaries into our imaginations, that quickens the heart and snatches a breath. Yet this wasn't some nightmare rising as if from the murky depths of the unconscious, but a benign form drifting dreamily into view: a basking shark. Then a second slightly sharper-ended fin appeared several metres behind – its tail – followed by a hint of its nose nudging the skin of the sea a couple of metres ahead: three points that gave an idea of its enormity.

Scars, bite marks and damage to dorsal fins, sometimes from boat propellers, can identify individual sharks, but Jackie and Graham didn't recognise this one, and everyone swung into action, ready to collect its essential 'passport' information. Jackie took the wheel and manoeuvred the boat so that Haley and Natasha could take photographs of its dorsal fin from both sides before we moved closer, carefully approaching from behind and to the right. I had been warned to watch from the far side of the boat.

'It may look peaceful,' Graham said, 'but when we get near, if it powers off and its tail sweeps out of the water you could be done for.'

He took up his position at the narrow prow and leaned over the railing lowering an underwater camera in front of the boat to film the shark in order to determine its sex – males have long claspers behind their pelvic fins and this was a female. So far it hadn't seemed aware of our presence at all, even though we were drawing close. At which point Graham picked up a long metal pole that half resembled a harpoon. Surely not!

No, this was the answer to the conundrum of how to collect DNA from an active basking shark.

'We have a mix of the expensive, like the boat and satellite tags,' said Jackie, 'and then we have the cheap and homemade, like this . . . '

Attached to the end of the five-foot pole was a short paint roller, to which a sterilised kitchen scouring pad had been fixed with two bulldog clips. Basking sharks are covered in a layer of dark mucous, and a quick rub with the pad on the dorsal fin is sufficient to pick up a decent slimy sample, so long as you can get close enough. They may be colossal, but basking sharks have brains no bigger than a cereal bar, with the largest portion devoted to smell and little to vision. Small eyes are positioned at the front of their wide heads, so by keeping out of the line of sight it is possible to edge to within touching distance when they are preoccupied with feeding.

Creeping up on a seven-metre shark is an unforgettable experience. As Jackie cut the engine and we drifted alongside, I could see its vast muscular mass in the water flexing rhythmically beneath towering fins, the sun illuminating its grey skin, marbled with black in patches along the back. It was so much stronger-looking than I had envisaged, and even longer than the boat. Ahead of us the wide pectoral fins were visible, as well as gaping gills sieving seawater that was being funnelled through its barrel mouth.

Graham stretched forward and deftly dabbed its fin with the scouring pad, at which point it caught sight of the vessel and powered down into the shadows with a surprising turn of speed, spraying us with saltwater as its tail swung from side to side. It's not often you get a refreshing facial courtesy of a shark.

The scouring pad was covered with fishy smelling gloop, and sections were cut and placed in alcohol-filled test tubes before the gear was prepared for a second encounter. The whole efficient information-gathering exercise had taken little more than a few minutes, and the shark, obviously unperturbed, was back at the surface and busy feeding once again.

Then something caught my eye further off in the sea, like a computer cursor crossing a slate-blue screen. 'There!' I said, and it was my turn to deliver the line: 'Sh-shark!' Okay, a bit clumsy, a bit hammy, but it did feel good to say it, and we made our way over.

This one was smaller: another female, though with noticeable white skin blemishes that could have been bite marks from an amorous male. We found ourselves slightly ahead of her, and as the boat turned it smoothed the water on the port side – a mesmerising moment when fins above the water became something more below and we could clearly see her trawling through the plankton with gill rakers open wide. At which point, a simply enormous shark came out of nowhere, fully submerged and following within a foot of her tail. Its mouth could almost have swallowed her whole. No one had even been aware it was there, and everyone gasped.

'That's huge!' Jackie exclaimed. 'The largest I've ever seen.'

Basking sharks have reportedly reached over ten metres, but at eight metres this decades-old male was about as massive as they come in our waters, and it headed right under us. Our craft suddenly felt very small indeed. Like a grey submarine, he glided beneath, close behind the female, and moved away hidden below the sea.

'Did you see it? Did you see it?' everyone was asking each other.

Not only was it an extraordinary sight, but its nose-to-tail behaviour was likely to indicate a breeding bond. It is believed that the sharks mate and give birth in the area, and a newborn, about the size of an adult seal, has been recorded close to the Isle of Man.

There was no time to dwell on what we had witnessed because another shark was spotted nearby and its details needed logging. I could see Graham's underwater footage relayed live on a screen in the cabin as we approached, showing first the crescent-shaped tail ahead, then eventually its solid body that could be mistaken for a great white shark were it not for its faintly bewildered look of open-mouthed astonishment as it fed, like an exaggerated OMG!

And the sharks just kept coming. By early afternoon the labelled test tubes of DNA swabs filled half of a foot-long storage tray, and Jackie was worried we were running out of alcohol – a common enough occurrence on boat trips, but not usually for scientific reasons. We located the large male once again, though he was cautious and dived whenever we came near, and the dorsal fins of sharks identified on previous trips were also noted. Telling them apart takes a lot of experience, however the crew reeled off names like they were familiar friends: 'Ah, there's Neve!' 'Groucho Manx has put in an appearance!' 'Over there, it's Dermot!'

After stopping off in Peel for a cup of tea at a harbour café, we got back to work, with me helping as much as I could by trying to keep out of the way, which isn't always easy on a small boat. The conditions got better and better, and at one stage, as we coasted south from Peel Castle, there were three basking sharks feeding behind us and one in front.

Eventually we came across a seven-metre male that Jackie and Graham considered suitable for tagging, and once all its vital details were recorded, Graham managed to prod the attachment in just behind the dorsal fin and off it swam. Satellite-tracking tags, which resemble a small torpedo with an antenna on top, store data about underwater movements and transmit a shark's position whenever it surfaces. They have confirmed that basking sharks are active in winter, and don't hibernate as once thought, as well as revealing the extent of their travels. Individuals tagged off Scotland's west coast have journeyed as far south as the Canary Islands, while one from the Isle of Man, 'Tracy the Tower Insurance Shark', voyaged right across the Atlantic. At £3,500 per tag, it seems a little commercial sponsorship goes a long way.

The shark we had tagged was named Fricassonce, a title conjured up by Natasha and Haley, and regular updates on his progress would be posted online. Publicity is a key part of Manx Basking Shark Watch's work because, as Jackie puts it, 'if the public don't know what they've got, they won't know what they're about to lose'.

Not only had my wanderings so far impressed on me the wealth of fauna we are lucky enough to have in Britain, but I was also beginning to realise that conservation is as much about people as it is about animals. I had met dedicated folk striving to make a difference, and generally earning far less than they should (note to jobseekers: destroying the planet pays a lot better than working to save it), and it seemed that, quite apart from trying to protect wildlife, they were also fighting to preserve and promote those values necessary to conserve the natural world. Compassion can be lost as easily as species, and when it goes, then plants and animals are sure to follow. It is not enough simply to bequeath biodiversity to future generations without also passing on a sense of its significance and, perhaps hardest of all, a genuine love of life on Earth. If our children don't give a monkey's about, well, monkeys, then the future is bleak for plenty besides.

Large, awe-inspiring creatures are a public-relations dream for conservationists. You are never going to get people of any age excited about zooplankton, but as ambassadors for healthy oceans, basking sharks are fairly irresistible. I had been lucky enough to join Jackie and Graham on their best day ever. At 9.30pm, we packed up the samples and headed back to port across Niarbyl Bay. The sun was sinking in a blaze of colour and sharks were still visible, their dorsal and tail fins moving from side to side in serene t'ai chi moves, unfurling ribbons of light behind them as they went. And as Graham steered us towards the harbour at the end of the exhilarating outing, which had included thirteen close encounters, I turned to him and said with a smile: 'Right, what else you got?'

I needed to catch my breath. Racing around the country, I had so far seen four out of the five scarce reptiles and amphibians on my list, three of the five endangered fish, four of the target mammals, three threatened birds and two rare invertebrates – in all, over half the twenty-five species I had originally set out to encounter. Only, a spell to relax and reflect was out of the question. It was mid-June, the wildlife wouldn't wait for me, and spare time was in short supply. One of my work-contract requirements was that I actually turn up, which was extremely inconvenient, while at home the chores multiplied to fill every free moment, with tasks I had persistently put off refusing to make life easy. When I tried to tackle the lawn, the electric mower conked out part way through. Why now? I cursed, as I tried and failed to restart it (then again, when else would a lawnmower break?). Our cat and dog ensured I never got a minute's peace, always wanting to be outside when they were inside or inside when they were outside, and keeping me busy clearing up puke, mud and dead mice as they waged their unremitting war against clean carpets. And a combination of gravity, rainwater and people touching things I had mended before served up almost daily additions to my DIY duties. Slates slipped, gutters leaked, drains blocked, fuses popped, light bulbs died, curtain rails slumped . . . I decided that our Victorian property, which should have been named Good Intentions, was old and simply wanted to lie down, and the kindest thing to do over the coming months was to let nature take its course – either that or we all tiptoe out, gently shutting the front door behind us, flee to a Travelodge and pretend home ownership had never happened. 'What, that house? Nope, never seen it before in my life.'

On top of all this, I had my eldest daughter's latest boyfriend to fret about. I had attempted the usual interrogation about where he lived and his career aspirations without getting far, before finally resorting to that old chestnut: 'And what do his parents do?'

'How should I know?'

'Well, haven't you noticed his mum or dad talking about being on call at the surgery, or perhaps preparing for a high-profile legal case the next day?'

'No.'

'Hmm. Ah well, I'm sure he's nice enough, in the short term. But don't go thinking these things have to, you know, last forever.'

So, with nothing really sorted at home, but everyone seemingly happy enough, I was off once more, this time for a few days in East Anglia, and, being a gentleman, I had actually invited my wife to join me if she wanted.

'No guarantees, but there could well be the chance of some rather exciting examples of Odonata, Arachnida and Ranidae,' I told her.

'Well, you sure know how to treat a lady,' she said, adding after a pause: 'I'll get your bags.'

I drove through the night, managing a couple of hours of sleep in a service station car park. By morning I reached Norfolk, where I met with Jim Scott, an RSPB site manager who also oversees the charity's protection scheme in East Anglia for our rarest breeding bird of prey: the Montagu's harrier.

Harriers are a group of medium-sized raptors that tend to hunt low and slow over open ground, and Montagu's harrier is the smallest of the three British species. It is also easily confused with the hen harrier, as the two males are grey with black wing tips and the females streaky brown with white rumps. At one time, both males were considered one species, the females another, and it took keen-eyed eighteenth-century naturalist George Montagu – him of Slavonian grebe fame – to sort out the muddle. The 'ash-coloured falcon', as it was then known, was subsequently renamed in his honour.

Even in Montagu's day, the species would not have been common, with perhaps a few dozen pairs dotted around heaths, marshes and moors. The UK lies at the edge of its range, and cold springs and wet summers do not particularly suit this ground-nesting bird, which spends the winter south of the Sahara. Hardly surprising, then, that with so few to spare, indiscriminate killing by gamekeepers during the nineteenth and early twentieth centuries, coupled with the nest-plundering activities of egg collectors, almost exterminated this striking harrier altogether. And what post-war revival it did enjoy was short-lived, rising to thirty pairs by the 1950s before the impact of agricultural intensification and introduction of pesticides, such as DDT, into the food chain knocked it back again. In 1974, the number of breeding pairs recorded hit zero.

Against the odds, this unpredictable migrant did stage a faltering comeback, and it is thought those that started breeding here once more were of French or Spanish origin, as they adopted the southern European habit of setting up territories on arable land. Today, nondescript fields of wheat and oilseed rape are the nesting locations of choice for our 'Monties'. Which is a bit awkward. We create all these lovely safe reserves, and they plonk down and start laying in the middle of some vast industrial-scale crop nearby.

'Fortunately the overwhelming majority of farmers are very helpful,' Jim said, as we chatted at his office at Titchwell Marsh RSPB reserve. 'Sometimes the young fledge before the field is harvested, though more often we have to mark the area around the nest with canes so it can be left uncut.'

The added danger of disturbance from overeager birdwatchers and photographers, and the risk of alerting egg collectors, means that the dozen or so nests detected every summer across southern England are generally kept secret and monitored full-time to ensure no harm comes to the brood.

'There's obviously a demand from birdwatchers who want to see one, and we're caught between wanting to protect them and wanting to promote them,' Jim added. 'So when they nest in suitable locations, which can happen every few years, we organise public viewing with landowners and that's been very successful. Lots of people keeping an eye on them can also prevent problems and relieve potential pressure on other breeding sites.'

Normally you get five to six pairs in East Anglia, but he told me that the year was shaping up to be one of the worst ever.

'How many pairs are there?' I asked.

'One.'

Oh.

However, he had arranged for me to visit the site. In order to get security clearance I had previously satisfied him that my only documented crimes against birds involved lame puns, and promised that when it came to Montagu's harriers I would avoid any tortuous jokes. The full Monty repertoire would be kept under lock and key. And, constricting as the agreement felt, I wasn't a snake in the grass prepared to squeeze anything more out of the name, like some Monty python. No, I was good for my word.

I was to meet RSPB harrier-protection warden Bob Image at a B-road turn-off in north Norfolk, and I arrived early and parked in a lay-by. Arable plains stretched into the distance on all sides, the wheat silver-green and the darker rapeseed turning yellow. It seemed a uniform and unwelcoming environment for nature, and yet there were plenty of small birds in the hawthorn hedgerows, skylarks were singing above the crops and I spotted a hare running along a field edge and dodging among a scattering of wild poppies. I was also pleased to see a bird of prey flying over a stand of trees further off – good to know they were about, even though this was a buzzard, heavier and broader-winged than a Montagu's harrier. A few minutes later, a slimmer raptor appeared: a red kite, much closer to the shape I was looking for. And then, taking me by surprise, a harrier came into view, gliding low over the far boundary of the field in front of me, with wings held up in a 'V' shape. Was this it? I desperately twiddled the focus wheel on my binoculars until I could make out its plumage. No, it was a marsh harrier, the largest of our three harriers, but a great sighting nonetheless. Just how many birds of prey were there in this place? Was some local falconry show being held that I wasn't aware of?

Bob pulled up in his car and I hopped in. We drove down a long track, scattering pheasants, to a grassy area, where he tucked the vehicle in behind a hedge and we set up our telescopes nearby: mine a cheap and fairly new one, his a trusty battered old field scope, which said something about the years he had been working outdoors. Aged in his sixties, Bob began his bird surveillance work in the early 1980s, working to protect goshawks and harriers, and now spends his summer days patiently observing nest sites.

But he keeps his distance. 'This pair's first nest failed, the eggs were probably eaten by a ground predator, and they're trying again in the crop cover just the other side of that hedgerow,' he said, pointing over two wide barley fields to a border of hawthorn about a kilometre away. No wonder we needed the telescopes.

'The male can travel a long distance to find food to bring to his mate on the nest, so we could have to wait hours to see anything,' he warned me.

We didn't have to. After just five minutes, one appeared, long-winged and pale grey, floating on the breeze above the ragged line of hawthorn.

'There!'

With the naked eye it would have been hard to make out much, but our powerful lenses came into their own, and it was a wonderful sight. Slender and graceful, he flew into the wind with effortless strokes, and I could clearly make out the black tips to his primary feathers and the dark line running along the back of his narrow wings that was a key distinguishing feature.

'You can see how light and buoyant he is in flight,' Bob said. 'There's nothing to them, they're all wings and feathers.'

Montagu's harriers weigh much the same as a box of cheese crackers, yet have a wingspan of over a metre, which means they can keep airborne at slow speeds as they quarter the ground in search of small birds and rodent prey.

The female, brown as a buzzard, then rose from the nest area, and they perched close to one another on the hedge as she preened. After which they just sat there for thirty minutes, looking around them as if wondering what to do next.

Their answer came out of the sky. A red kite, perhaps the one I saw earlier, coasted high over the field in front of us, a little too close to the nest site, and the male Montagu's harrier took to the air to defend his territory. You couldn't ask for two more attractive and agile birds of prey on the wing together, and Bob was equally enthralled by the sight.

'Quite something to have a Montagu's harrier and a red kite in the same viewfinder,' he smiled, as we watched through our telescopes.

Lankier and smaller he may have been, but field marshal Monty on the offensive was not to be trifled with, and the red kite was swiftly harried from the vicinity. The male Montagu's harrier sailed off, the female returned to her hidden nest and, matinee show over with little prospect of much happening for a while, we drove back to the main road, where I said goodbye to Bob, thanking him profusely. To see one Montagu's harrier is certainly special, so a pair was more than I could have hoped for, and I was on a high throughout the journey to my next destination – another top-secret location.

I was equipped with a map of Norfolk marked with a cross, a new pair of wellington boots and a name written on a piece of paper. The man I was to meet was ecological surveyor John Baker, the place was confidential and the unused boots were a biosecurity precaution to ensure I didn't bring any diseases onto the site. Wiping out a decade-long project on a single visit wouldn't have done much for my conservation credentials.

It took a little while to find my way, but eventually I located the spot, branching off a lane and following a muddy track through an area of woodland to reach a concealed clearing. The afternoon was warming up and I could hear cuckoos calling from beyond the trees. John arrived soon after and, sporting my shiny new wellies, I followed him between stands of silver birch and willow to an open area pockmarked by round pools – far larger than the bomb-crater ponds I had come across on my smooth snake search in Dorset, though equally neat and circular.

'Pingos.'

Beg your pardon?

Pingo is the Inuit word for a small hill (not a cartoon penguin) formed when a block of ice gets buried in earth. As the ice melts, the soil collapses, leaving a shallow depression that can fill with water, and at the end of the last ice age areas of East Anglia were riddled with these geological features. Most have grown over or been ploughed up, but some still survive, and those that are carefully maintained make ideal homes for amphibians – in particular the species I had come to see.

John guided me through a stretch of ferns and sedge to one of the larger pools. Round and shallow with a defined rim – a profile much like an upside-down frisbee – the woodland oasis was thick with vegetation and buzzing with insect life. Yellow flag iris, lady's smock and water violet flowered at the edge, and the surface was covered in a profusion of broadleaved pondweed. We walked slowly, checking the banks ahead for signs of movement, and came to a stop halfway round. Reaching into his bag, John pulled out a CD player and switched on a strange guttural noise, which sounded like a cat purring and retching at the same time. Repeated on a rolling loop, the recording took on a mechanical quality, unlike any animal I had ever heard before. But it did the trick. Something replied from the centre of the pond: a couple of squelchy croaks at first, then a few more in succession. If the call of the natterjack toad comes out of a toy shop, then this was more from the factory floor – a motorised pumping rhythm.

'Can you see it?' John asked. 'About five metres out and level with the tree stump.'

I couldn't at first, and it had fallen silent.

'He's still there,' John added, pointing me in the right direction.

Ah! In among the green-and-brown pondweed I could make out a green-and-brown face peering back at us – an endearing froggy face.

Those readers taking notes while reading this book – the majority, I assume – will have noticed that I referred much earlier to Britain having just six native amphibians: the palmate newt, smooth newt, great crested newt, common frog, common toad and natterjack toad. Herpetologists who didn't throw the book away in disgust at that oversimplification can stop their tutting and grumbling now and breathe a sigh of relief. Their perseverance has been rewarded. What I was staring at was indeed our seventh species: the pool frog.

The pool frog? To be honest I hadn't heard of a pool frog until I began drawing up my list of UK rarities. But that is fairly forgivable, given that it was not considered a resident species until recently, by which time it was too late and the last few had gone from our countryside.

Everyone always supposed that any pool frogs found living in Britain had been released by enthusiasts or bored collectors. Deliberate introductions of non-native amphibians and reptiles have been going on since the 1800s, and you can find midwife toads, alpine newts, wall lizards and even terrapins trying to make a go of things this side of the Channel. But the pool frog was different: it had been here all along. And it took buried bones, archive tape recordings, historical accounts, DNA samples and years of painstaking detective work to finally establish what we had lost.

About the same time as the pingos were forming, pool frogs joined a host of animals hopping, slithering, crawling and flying north in a race to colonise exposed land after the glaciers retreated and before meltwaters cut off Britain and much of Scandinavia. Similar to common frogs, but more strikingly patterned with dark spots on the flanks, a light line down the back and a white vocal sac, pool frogs made their home in eastern England. We know this because their bones have been unearthed in excavations of Saxon settlements in Lincolnshire and Cambridgeshire. However, the draining of the Fens and later changes in land management meant they became scarce and localised. Archival evidence from the 1760s, glass-eyed museum specimens from the mid-nineteenth century, hundred-year-old illustrations and old photographs prove that they clung on, though in low numbers. Only this wasn't to last. By 1995, Britain's pool frogs had died out in the wild, with one surviving in captivity until 1999.

Once it became clear that we had let a native species slip away virtually unnoticed from under our noses, a decision was made by Natural England to reintroduce them. Such a cut-and-paste approach to conservation can prove controversial, as it has been with white-tailed eagles, red kites and beavers, but the benign pool frog was deemed an ideal candidate. There were plenty for the taking on the Continent, except that our former pool frogs, whose calls apparently had a distinct Norfolk 'accent', were found to be more closely related to a northern race of threatened relatives in Scandinavia than greener-looking counterparts in mainland Europe. So Sweden was asked whether it would mind sparing a few of its precious adults, along with some tadpoles and spawn, and kindly agreed. These were flown over in four annual batches from 2005 and released in the carefully prepared pingo ponds, with the site kept secret to prevent potential thefts or interference, and since then they have bred successfully. What I was staring at was one of around three dozen descendants – a rare, British-born pool frog.

These active amphibians have plenty of character, John told me, and just like holidaymakers they demand sunshine and a pool. 'The hotter they are, the happier they are,' he said. 'They like to bask out of the water and dive in if disturbed, and they're also great to work with as they are not timid at all.'

An insect moving beside a neighbouring pond coaxed a curious male to within a couple of feet of the edge, and he didn't seem the slightest bit worried by us approaching. He was a handsome fellow with big eyes breaking the surface, and as he crawled over the weeds his markings were clearly visible. Seams ran down either side of his brown back, parallel to the central pale line, and a white throat emphasised his wide mouth. He gave us a hard stare, as if we had just put a towel down on his sun lounger. Then a female with much more defined colours moved into the shallows, and John, who works on behalf of Amphibian and Reptile Conservation, took plenty of photos because each of the frogs can be identified by its individual markings.

'There isn't an expectation they will become a widespread species, and I suspect they never were, but the hope is that they can be restored within their known natural range,' he said.

Despite appearing laidback, the frogs are actually quite fussy and need ponds kept in prime condition, so local reserves are considered the most suitable option for future introductions, he added. However, building up sufficient numbers is a time-consuming task, especially given the impact of poor weather and predators, and John runs a makeshift maternity ward at his home, consisting of rows of bowls and trays in which pool-frog tadpoles collected from the ponds are given a head start before being put back.

'As you can see, they're well worth the effort,' he said.

I couldn't agree more. Good to know these splishy-splashy, lilypad-leaping, sun-loving frogs are back.

I had two further species of wet places to see while in East Anglia, and spent that night in the suitably watery setting of Wroxham on the Norfolk Broads at a riverside hotel. In keeping with the damp theme, it also started to rain, which was a worry.

A clean and comfortable room was a joy after the previous night in the car, and in the morning I ate breakfast overlooking the river Bure. Rows of dinghies were lined up neatly by grass banks, and the homes opposite, so low to the water, looked as though they had also been moored between the weeping willows. It was a peaceful scene and I had a bit of time to read my guide to dragonflies – once I had finished perusing the breakfast menu, which valiantly extolled the health benefits of the full English fry-up, noting the protein goodness in sausages, antioxidants in fried tomatoes, the 'slow-release energy' of baked beans and that kippers boosted 'creativity and concentration'. I had the kippers.

My rendezvous point was a few miles downriver at Horning. According to the village sign, the name means 'the folk who live on the high ground between the rivers' – impressive for such a short word – and I waited in my car beside the main road for Pam Taylor to turn up, following her in convoy until we reached Alderfen Broad nearby. The good news was that Pam got out of her vehicle wearing a T-shirt emblazoned with a large illustration of a golden-ringed dragonfly. Not because of the design, encouraging though it was, but because it was T-shirt weather. Dragonflies won't take to the wing if it is too cold, raining hard or very windy, and it looked like the conditions were going to be kinder than forecast.

Managed by Norfolk Wildlife Trust, the secluded Alderfen Broad nature reserve consists of grassy bogs and woodland surrounding a large lake that was originally created by medieval peat digging, in common with most of the Broads. It provides an ideal habitat for dragonflies and, according to Pam, was an excellent place to find the national rarity we were after: the Norfolk hawker.

Pam has dedicated three decades to working with dragonflies, rising through the ranks of the British Dragonfly Society to the post of president, and has seen everything from the enormous 'helicopter' damselflies of Central America to scarce migrants and eye-catching British beauties – and she has the T-shirt to prove it. But she puts the Norfolk hawker at the top of her list of favourites.

What makes this species special is not just its scarcity, its aerial mastery or its dashing good looks, with a long gingery brown body and apple-green eyes, but also the fact that it is paying back all the hard work put into its conservation. Lost from many wetlands due to drainage, pollution and the neglect of water-filled ditches, and confined to a handful of fens and grazing marshes in Norfolk and Suffolk, this endangered dragonfly has been gradually strengthening populations in former haunts as water quality has improved and habitat has been restored. On the day I met Pam, information was emerging that a new breeding site had been confirmed in Cambridgeshire – the first in 120 years. Welcome news, given that we have lost two species of dragonfly in the past half century and almost a third are declining or considered threatened.

Fortunately more and more people are taking an interest in dragonflies, with nature reserves actively promoting them. They are without doubt incredible insects, but they can also be a little alarming. Some airborne critters send us into a flap because of their poor steering, like May bugs and moths, their outright aggression – horseflies and midges included – or potential to cause pain, such as wasps and bees. Dragonflies share none of these characteristics. Sizable, fast predators they may be, darting around like darning needles, but they are in total control in the air, mean us no harm and, thankfully, can't sting. It is their large, fencing-mask eyes that can be so disconcerting. They watch us like no other insect and don't seem half as afraid as we would like them to be. Unnervingly confident in flight, a hovering dragonfly can halt you in your tracks, scrutinising you like a miniature spy craft before zipping away out of sight – doubtless plugging into a secret docking station in order to download its footage. Who exactly is operating these creatures?

Not all species are alike though. The fifty-plus members of the Odonata order found in the UK include dainty damselflies: thin straws of colour best recognised by their weak flight, the way that they hold their wings above the body at rest and the position of their eyes, set apart on a dumb-bell-shaped head. Quite exquisite, they include the gregarious common blue damselfly and widespread large red damselfly, as well as the iridescent banded demoiselle, which has a thumb mark of black on its wings.

The more familiar true dragonflies are strong fliers, with eyes close together and wings spread wide when stationary. They include the stout-looking chasers and skimmers that wait on lookout perches before darting out after flying prey, and hawkers, which ceaselessly patrol territorial airspace, snatching in-flight meals using their barbed legs. The hawkers are arguably the most magnificent of the bunch, and boast the emperor dragonfly among their ranks. This bulky beast, sporting a green body and, in the males at least, a striking blue abdomen, even hunts other dragonflies, snatching them from the skies and devouring them in mid-air. If you find that scary, imagine taking a stroll through the swamps of the Carboniferous period three hundred million years ago, when bird-sized dragonflies with two-foot wingspans roamed the riverbanks. Yikes.

We hadn't gone far up the path into the reserve when Pam pointed out a four-spotted chaser on low vegetation in a patch of sunlight. It was one of about three species I could actually recognise, so I nodded knowingly: 'Ah, the old four-spot.' A widespread and common dragonfly, it is lovely looking. The body is golden brown and this colour washes slightly across the front of its wings, which are marked with the telltale spots. They are also easy to observe because they often return to the same perch after territorial skirmishes and hunting forays.

'And there's a black-tailed skimmer,' Pam said, gesturing towards a yellowish dragonfly of a similar shape settling on the path – another common species, and one I wanted to remember. 'And that's a male,' she added when we came across a lighter blue version further ahead.

Fortunately the names of our dragonflies are fairly straightforward to learn. Unlike the majority of insects, which are known only by formal taxonomic titles, they have managed to charm their way out of the mire of Greek and Latin into everyday English. Given that this has come about relatively recently, these everyday terms lack many of the idiosyncrasies and cultural connections that have evolved over time, as in the case of, say, British birds. However, the simple descriptive labels still have an undeniable appeal, from the azure damselfly and beautiful demoiselle to the golden-ringed dragonfly and brilliant emerald. Who wouldn't want to see something called a brilliant emerald?

The next species we chanced upon didn't have such an alluring name: the hairy dragonfly. On close inspection, its thorax did look a little frizzy, but that aside it was extremely attractive. Almost totally clear wings were broken up by veins, like cracks in a windscreen, and the long, dark abdomen was ornately patterned with dots and dashes of pale blue. It took off, stiff wings rustling through the reeds, and I noticed a female at the edge of a slow-moving water channel dipping her tail beneath the surface to lay eggs. Like other dragonfly eggs, these would hatch into ferocious aquatic nymphs, also known as larvae, which live for a year or more and eat anything smaller than themselves that they can grab with their extendable jaws. When they are eventually ready to undergo the final stage of metamorphosis, nymphs crawl out of the water up a plant stem and, skipping the pupa stage altogether, the outer skin splits and a dragonfly emerges, pumps itself up to full size and heads away from the breeding ponds for several days to feed and mature before returning with colours blazing to engage in the hurly-burly of courtship.

'Their short lives as flying adults are fast and furious, and generally last only a few weeks,' Pam said. 'Very few die of old age. Predators such as birds eat them, they starve in bad weather or get damaged during mating and in territorial battles, which means they can't fly well enough to hunt.'

We spotted a pair of hairy dragonflies settled on a fern and linked together in the latter stages of mating like bent coat hangers. Male dragonflies transfer sperm from their hind end to the front of their abdomen before grabbing a passing female by the scruff of the neck using tail claspers, designed in a range of spanner sizes specifically to fit their own species. The female curls her abdomen around to collect the sperm, in what is known as a copulation wheel, then heads off to lay eggs, sometimes guarded by the possessive mate or guided in his clutches.

We followed the path to a corner by the lake, and wherever we looked there were dragons and damsels. The air was threaded with colour, and every other reed and grass stem seemed to have something bright clinging to it. Pam put names to species as we passed, and I worked hard to absorb the variety of life around me: a blue-tailed damselfly that looked as if it had been dipped in paint; a variable damselfly with a bar code of black stripes along the abdomen; pretty azure damselflies; a red-eyed damselfly; another four-spotted chaser; large red damselflies resembling biro refills, and plenty of black-tailed skimmers. And then she pointed at something moving fast overhead, tracing wide geometrical shapes in front of a stand of overhanging trees.

'There we go,' she said. 'That's a Norfolk hawker.'

Identifying dragonflies in flight takes some doing. You have to learn how to decipher fleeting information, and the faster you can read the visual cues the slower the insects appear – a passing blur of wings becomes a size, a shape, a pattern . . . one possibility rather than another. With Pam's guidance I was able to discern diagnostic features: a caramel brown body, colourless wings and green eyes. Prized apart and put back together, the elements all added up to the rarity I had come to see, and distinguished it from the similar and more common brown hawker, which has bronze-tinted wings and eyes and flies later in the summer. This was a Norfolk hawker, powering through the air like it owned the place.

For hawkers they do alight quite frequently, fairly low down in foliage, and the challenge was to find one that had settled. An area had been cleared of trees, and Pam led the way through the high reeds and grasses that were bristling with various dragonflies until, using binoculars, she spotted a Norfolk Hawker clinging halfway up a stem. Stealth was required to get close enough to see one of its signature marks: a yellow triangle at the top of its abdomen from which it gets the second part of its scientific name, *Aeshna isosceles*. One step too many and off it flew, though I managed to track another flying into the cover and discovered it was not alone. For such a frenetic insect, what I came across made a surprisingly tranquil picture: two rare Norfolk hawkers resting alongside each other amid the swaying green reeds, wing tips almost touching.

Leaving Alderfen Broad, we drove to nearby Ludham-Potter Heigham Marshes, a national nature reserve where Norfolk hawkers are known to breed in the narrow dykes that cut through buttercup-filled pasture. The ditch by the entrance was almost choked by a plant called water soldier, which is known to be important for the dragonfly. Its floating rosettes of serrated spear-shaped leaves, looking a bit like pineapple tops, not only support invertebrate prey but also provide a good grip for emerging nymphs, and Pam wanted to search for the cases left behind, called exuviae. Walking along the bank, we found the sizeable dried husk of an emperor dragonfly nymph still attached to a water soldier leaf, and soon after a slightly smaller casing just within reach. A quick check with a magnifying glass confirmed it was that of a Norfolk hawker, exactly what Pam was after and another important addition to the records. It also illustrated how getting the conditions right, even with something as simple as a water-filled trench, can prove a lifesaver when it comes to helping endangered species.

I had more ditches to explore, though the next species comes with a warning. If you're not a fan of spiders, then now is the time to look away.

(And they don't come any bigger than the one I was hoping to find.)

Considering we have over 650 kinds of spider in Britain, the vast majority of us are clueless about all but a handful of species. Most people will know the large house spider – that fast-moving dark brown beastie whose main habitat seems to be bathtubs and skirting boards. Then there is the common garden spider, which builds webs in hedges and is marked with a dotted white cross on its fat abdomen. Who could forget the knobbly kneed daddy-long-legs spider, with its messy webs in corners of rooms, and the zebra spider, which can be found hunting along outside walls and pouncing on prey. Oh, and tiny money spiders, of which there are many varieties, familiar for using strands of silk to carry them on the breeze. And then there is . . . hmmm. Well, that's enough to be getting on with. Anyone able to name five more can give themselves a pat on the back. Spiders are not exactly the easiest subject to master. The colloquial names rapidly run out, and beginners face a daunting diversity of small, similar-looking species. In some cases it takes a detailed examination of their genitalia under a microscope to tell spiders apart and, fascinating as animals are, I do draw the line at such matters.

There is no denying that a large proportion of people like to keep their distance from spiders, genitalia and all. Harmless ours may be, and useful predators of pests, but an irrational fear of them is commonplace. Not so for Helen Smith, whose kitchen has been known to house thousands of baby spiders as part of a breeding programme for one of our most endangered and spectacular species: the fen raft spider. With an outstretched span as wide as a digestive biscuit, this remarkable semi-aquatic spider is big enough to catch sticklebacks and is handsome, too, adorned with two cream stripes running down its dark brown body. You won't find one creeping over the linoleum or scuttling under the dresser though, because they spend their lives lurking beside still and slow-moving water, waxy haired feet dimpling the surface as they wait patiently for vibrations to alert them to possible prey.

Helen, who is president of the British Arachnological Society, has dedicated the last twenty years to conserving this charismatic rarity, and I was able to join her at Suffolk Wildlife Trust's Redgrave and Lopham Fen reserve, one of three established UK sites for the protected species. It was here, in the alkaline pools of this spring-fed wetland, where the fen raft spider – as opposed to the more widespread raft spider of acidic bogs – was first identified in Britain in 1956. However, it faced a battle for survival during the decades that followed its discovery. A borehole drilled to supply local homes with water drained the chalk aquifer, leaving the spiders high and dry, and only a few dozen adults managed to make it through the hot summers, confined to a handful of ponds. Eventually the pipeline was moved and the valley restored to its former soggy glory. Scrub was cleared, new pools dug in the peat, grazing was introduced and the eight-legged residents have become something of a star attraction, with their own visitor trail and pond-side viewing platform.

This was where we began our search, standing on the short wooden boards and peering into a weedy circle of water a few yards across that was set among extensive beds of great fen-sedge in the centre of the reserve. A warm breeze was ruffling the surface, which Helen said made it difficult for the spiders to detect subtle vibrations and could jeopardise our chances of getting lucky.

'In the wind they retreat into the margins, so they're much harder to spot,' she said, scanning the edges of the pool with binoculars. 'Windy days can be a lost cause and we may not actually be able to . . . ' She paused. 'Ah, there we go.'

'What? Already? Where?' I wasn't expecting such instant success.

With guidance I was able to pinpoint what amounted to a couple of legs sticking out from behind stiff stems of emerging vegetation at the far side. It was amazing that Helen had noticed anything at all, but on moving position I could see it clearly. This was a female, the largest of the sexes, waiting out of the breeze, totally motionless, with one hind foot anchored on a bent leaf and the others splayed out on the water, pressed into the liquid light of surface reflections. She was certainly large, and the pair of stripes, like white chocolate drizzled across her dark, velvety body, were extremely smart. The shape of her abdomen indicated she was also heavy with eggs, Helen told me. These would soon be laid in a silken egg sac and carried beneath her for three to four weeks, kept moist in dry weather by being dipped in the water. When ready to hatch, the female would construct a tent-shaped nursery web high up in the sedge and guard the little spiderlings until they dispersed a week or so later.

'When people learn more about spiders they realise not only how little they need to fear them, but also how fascinating their life cycles are, especially their strong maternal instinct,' Helen said.

She has regularly turned arachnid crèche leader herself, hand-raising young fen raft spiders in her home in individual test tubes filled with damp cotton wool and feeding them on flies caught around dung heaps and compost piles. Part of a successful project to introduce the spiders from Redgrave and Lopham Fen and Sussex's Pevensey Levels to other suitable reserves, this labour of love has garnered plenty of press coverage. The spider-woman headlines just write themselves.

'There are numerous rare invertebrates that are small and inconspicuous and will never get much attention,' she said. 'So it's important we have popular flagship animals like this that help us conserve important habitats for a whole range of species.'

Helen needed to catch one as part of her licensed work, and we walked on between two lines of small round ponds laid out like bootlace holes on either side of the path. She had brought along chest-high waders for the job, telling me that her favourite larger pair, which she referred to as her 'maternity waders', had been nibbled through by mice after serving her well down the years, even during pregnancy. Pulling on the new pair, as well as gloves to protect against the razor-sharp sedge, she lowered herself into one of the deep pools and began searching between the overhanging foliage at the margins. 'There are lots of places where they can hide in the dark flooded caverns that go back a long way,' she said, heading deeper into the swamp.

Hoping to come across one myself, I looked in neighbouring pools and a short while later caught a glimpse of a large spider retreating rapidly from the water's edge. I called out to alert Helen, only there was no reply. Returning along the path, I checked the area where she had been but could not find her, just an empty, uninviting bog, capable of swallowing people whole, and acres of high sedge stretching away on all sides. It seemed a similar scenario to the time when Sarah Henshall disappeared in long grass on the bombardier beetle hunt. To lose one ecologist might be regarded as a misfortune; to lose two looks like carelessness. Then I heard a voice from the reeds and was relieved when Helen appeared. She was holding a small fen raft spider, one of last year's babies with another year to go before it would breed and die, and after releasing it we searched other areas until she spotted a nearly fully grown female basking in clear view, in a pool that had been dug only recently, which was a promising sign that they might gradually be spreading. Stealthily wading in, and passing a large leech on her way, she managed to get close enough to catch the spider quickly in her hands as it crawled down a stem in an attempt to hide underwater.

And what a catch it was. Even if you don't like spiders, you couldn't fail to be impressed by its size and markings. It was placed in a plastic beaker so I was able to get a close-up view, and there was nothing the least bit scary about it. Not that I was going to get all touchy-feely. It was fine where it was, safely in the container, looking up at me with its eight eyes . . .

This far from incy-wincy spider was a survivor: a good news story with legs. Almost wiped out when the land was sapped of water, its kind had made it through the tough times and, like the Norfolk hawker dragonfly, the species was slowly turning a corner. I was fortunate indeed to have met such a national rarity at close quarters, and spent a little longer scanning the reserve's pools for others, without luck, until it was time to leave Helen to her research.

Arachnophobia sufferers can resume reading once again.

Then, as I said goodbye, a giant hairy spider suddenly appeared from the shadows behind her and, like a monstrous bony claw rising up from the marshes, it leapt with fangs bared and . . .

Sorry, my mistake, you can resume reading now.

I had enjoyed a run of unexpected success in East Anglia, which was in some ways a shame because it meant I wouldn't be coming back in the near future. Tempted as I was to spend another day trying to find the rarities again on my own, I had to head home. One or two sightings is never enough, though I knew I would return some day, revisit the reserves, and perhaps have the opportunity to point out the dragonfly and spider to someone else – whether they liked it or not.

# Ten

How unsettling to hear your every breath, the repetitive passage of air that keeps you alive amplified within the wheezing windpipe of a snorkel. That rasping respirator noise resonates through the plastic tube, the bones of your ears, between your teeth, inside your skull. When you first push forward and the cold sea bites, every gasp of oxygen sucked down the cylinder is almost deafening. Inhale . . . exhale . . . inhale . . . exhale . . . never has the basic process of living seemed so loud. Gradually the urgency subsides. The rush of gases takes on a steady cadence, a syncopated rhythm in tempo with the dull bass beat of your heart and the rocking motion of your body as each flipper rises and falls. You can hear yourself relaxing.

It is impossible for me to ignore the sound of my breathing altogether, though. It remains a persistent reminder of why we don't belong in water. Face down, with the snorkel sticking upright behind my head, I constantly fuss with this thin teat that connects me to the air above, straightening it, adjusting the mouthpiece, spitting salty droplets clear whenever the intake begins to rattle like it has developed bronchitis. I can never trust that the open end is sufficiently clear of the surface, and worry about taking in a lungful of brine. Yet the greatest danger in this calm and shallow bay where I'm swimming is not so much the sea itself as the boats in the area. Low in the water, I am vulnerable to being run over, and pause often to look around me.

I have travelled only a short distance from the shore. Refracted sunlight plays over the gently shelving seabed and I spot a couple of shrimps on the sand a few feet beneath me. I pass through warm patches of water and cold, and stop to rinse out my face mask, lifting my head above the surface and enjoying a break from the sound of breathing as my emptying ears are flooded with the cries of seagulls and voices of early morning dog walkers on the beach where I left a towel and changed into my wetsuit. I pull my mask back into position, tight straps tugging at my hair, the suction of the oval rubber surround like a black anemone kiss fastening onto my face. It cuts out peripheral vision, so I turn my head as much as possible to take in my surroundings: a wide expanse of dimpled sand above which fish occasionally pass by, looking as if they're in a hurry to get somewhere.

Before long, I'm just out of my depth, and whenever I pause to tread water my flippers fan up glittering clouds of sediment, obscuring whatever I'm trying to observe. It takes practice to trust in the buoyancy of a wetsuit and come to a smooth halt. I haven't been snorkelling often and always feel like a clumsy imposter in some nature documentary, which is perfectly understandable given that our views of life beneath the waves are invariably through the window of television. The underwater world, as I know it, comes with a commentary – so I provide a little of my own to keep myself company.

'Well, here I am in Studland Bay on the Dorset coast, and what better place to be on a sunny summer morning.' Embarrassing, and my voice sounds odd through the tube, with my nose pinched by the face mask, but I persevere. 'If I lift my head from the water I can see the arc of chalk cliffs to the south that shelter this stretch of water. The headland, topped with fields and heaths, ends in the white stacks known as the Old Harry Rocks at Handfast Point.

'Behind me is South Beach, where I started out, lined with dark wooden huts set beneath a wooded hillside and the village of Studland a little way inland. To the north is the popular Knoll Beach and beyond that, hidden from view, are Bournemouth and Poole, and you can see in the distance a cross-Channel ferry heading off from there to France. Closer to me, several yachts are anchored in the bay, and I have to watch out for any on the move, but no one seems up and about yet, and as . . .'

I break off to cough after taking in a mouthful of salty seawater.

'Apologies, a slight technical hitch. As you can see, there's scarcely a breath of wind and only a slight swell, and when I dip my mask back beneath the surface . . . there we go . . . we get a wonderful view of what lies below, though the seabed is now over six foot down and looking a little out of focus, but you should still be able to make out that shell and that stone – not that we're out here looking for shells and stones.'

I decide to shut up and swim on. Up ahead I can discern what looks like the shadow of a giant cloud on the bottom and this is what I am aiming for: the seagrass beds. A few kicks of the flippers and blank sand gives way to a marine meadow that stretches away from me, lush and healthy. The sea floor is completely covered with a dense growth of thin green ribbons about a foot or two long. This is common eelgrass, a plant that can be found in patches in the clear and sunny waters of sheltered bays and inlets dotted around the coast of Britain. It spreads by dispersing seeds, like typical land-based grasses, and through its buried network of rhizomes, which bind together the seabed sediment, helping to protect sandy coasts from erosion. Nationally scarce, and internationally threatened, seagrass beds are home to species such as bream, bass and cuttlefish, and act as a nursery for fry and larvae that can hide amid the hairy tangle of narrow leaves.

I grab a drifting piece, translucent and so fresh and green it looks good enough to eat, and press its flat, slippery surface between my fingers. In the colder months, these fronds die back and the decaying brown strands get heaped up on the shore in stinking mounds, then, as the days lengthen, the matted rootstock sprouts fresh verdant growth. The slender leaves that I can touch with my flippers have reached full length.

Swimming slowly, I skirt the ragged periphery of the seagrass. Visibility is good, only I need it to be perfect. The distance from the seabed to the surface is no more than the carpet-to-ceiling height of a typical room, but this is a long way if the water is at all opaque. Winds from the east got into Studland Bay the previous day and churned up a little sediment, and in places the bottom remains distinctly murky, which is frustrating because the creature I wish to find is highly camouflaged and typically remains concealed low amid the stems. I need to be swimming down at floor level.

'Can you dive?' was the first question Neil Garrick-Maidment asked when I met him much earlier in the year.

I can't, and this obviously presents a problem. I am aware just how much it stacks the odds against me, stranded at the surface staring at the blurred green below in the vain hope that I might stumble across one of the rarest and most surreal animals on my list. When it comes to finding scarce species you make your own luck – though this time I need it by the bucket load.

There are shallower areas where the sun cuts through clear water and illuminates the sand and weed, and I focus my efforts on these, even holding my breath and plunging down for a closer inspection before returning to the top, spluttering and spitting and shaking the water from my snorkel. At one with my surroundings I am not.

What I'm looking for is most easily spotted along the edge of the seagrass, rather than in the thick of it, and I trace the beds south across the bay, then turn and begin covering the same ground again, scattering a shoal of a few hundred sand eels and spotting a big bass disappearing into the blue. Strange to think that at this time on a Saturday I would normally be trying to have a weekend lie-in, yet here I am, wide awake in a dreamlike world, floating among silver fish and swaying eelgrass. Even if I don't see what I have come for, it has been well worth the early alarm call and three-hour drive.

Quickening my movements to warm up, I chance upon an exciting find suspended motionless in an open area within the mesh of fronds: a pipefish. I'm not sure exactly what kind, but it is at least a foot long and marked with rings of brown and beige – a greater pipefish perhaps (it's the only species name I know, so let's hope so). As I move in closer my flippers waft the bottom and it disappears in a cloud of sand.

I have been in the water for almost two hours, and what started out as refreshing is beginning to feel cold, though I'm not ready to give up just yet. I manage another stretch of seagrass and there are plenty of fish about, including a second, smaller pipefish. Roll up its tail and it could pass for the closely related species I am after, which I'm beginning to realise is probably a lost cause. I know they are hiding down there somewhere beyond my field of vision – four have been spotted so far over the summer – but in anything other than gin-clear conditions I don't really stand a chance.

The beach is starting to fill up and boats are coming and going, their whirring propellers audible beneath the water. I decide it is time to head for shore, but as I stretch out with firm kicks my cold calf muscles immediately cramp up and I clutch at my legs in agony. I've never known anything like it. If anyone was watching me from the beach – groaning, with my knees up against my chest and arms flailing – they would rightly recognise that I am in trouble. Clenched up in a ball I'm not going anywhere for the time being, and I lie on my back, thankful that the foam wetsuit helps keep me afloat. I know I must force myself to straighten out, though flexing against the pain feels unnatural, like steering a car into a skid, and I have to grit my teeth and push myself until I can sense the stabbing spasms easing and am able to turn onto my front and paddle myself into shallower water.

I sit for a while among the waves, massaging my calves, and eventually get to my feet, strip off the snorkelling gear and clammy wetsuit and stagger stiff-legged along the tideline. As exits from the sea go, this is not exactly up there with Daniel Craig emerging from the waves in *Casino Royale*, and a face-mask ring around my eyes doesn't enhance the spectacle. Pride dented, but relieved to have made it back, I pick my way between families and finally slump down on my towel to dry out in the sun and rest my sore limbs. It takes a cup of tea from the beach cafe before I feel myself again. Then, with sand in my hair, fingertips wrinkled, neck sunburnt and ears clogged with saltwater, I return to my car and drive home, aching and exhausted, but strangely exhilarated. Minor drama aside, I have learnt a lot about the special marine life of Studland Bay and braved a few fears by exploring the sea alone. When it comes to the wildlife challenge ahead, I'm more determined than ever.

A baby seahorse looks a bit like a half-unfolded paperclip with eyes. It is tiny, irresistibly cute and extremely hard to keep alive in captivity, yet somehow Neil Garrick-Maidment developed the knack. As a teenager, he helped out at an aquarium that found itself with an excess of newborn seahorses in need of care, and he proved surprisingly successful at rearing the offspring at home on a diet of tiny live shrimps. It seemed inevitable, even from an early age, that he would become a naturalist – his first words, he tells me, were 'giraffe' and 'zebra' (not that he was raised in a zoo, though he did later work in one). Growing up in Malta, he volunteered at a dolphinarium and, as a school leaver, at a sea-life attraction when his family returned to England. Eventually he built a career around the care and conservation of species such as tigers, lions, wolves, even the odd giraffe and zebra, and became involved in wildlife filmmaking. But his early interest in seahorses endured, and his house was always crammed full of tanks housing a variety of specimens – enough of them to open the world's first seahorse aquarium at Exeter Quay, which attracted tens of thousands of visitors during the few years it ran before the exhibits were transferred to the newly constructed National Marine Aquarium in nearby Plymouth.

Today Neil continues to dedicate his working life to the welfare of these weird and wonderful creatures. As executive director of the Seahorse Trust, he energetically champions their cause, advising on breeding programmes and campaigning to stem worrying declines and safeguard fragile habitats.

We first met in early spring at a cafe in the grounds of the West Country stately home where the trust is based, appropriately for seahorses in a former stable block. Fit-looking, in his fifties, with a goatee beard and wearing a black-and-white desert scarf around his neck, Neil talked enthusiastically about the complicated biology and behaviour of these marine oddities.

'The more you learn about them, the more you discover just how fascinating they are,' he said.

Seahorses are like nothing else on the planet. No, that's untrue: they are a bit like everything else on the planet.

They look to have been assembled from an assortment of animal parts, possessing an elegant equine head, a long, anteater-style snout, chameleon eyes able to move independently, colour-changing skin stretched over a rigid, prawn-like exoskeleton, a kangaroo pouch for bearing young, and a prehensile monkey tail, which enables them to hold fast onto weeds in strong currents that would otherwise sweep them away.

'They are the most un-fish-like fishes – even people who don't like fish like seahorses,' Neil said. 'They have such endearing qualities. There's an engaging look about them that makes you wonder what it's like to inhabit their world and what they're thinking.'

Despite their confusing form and poor swimming abilities, seahorses do have undeniable poise. Upright and serene, few animals look more chilled out. And it is not just their unique appearance, but also their unconventional approach to breeding, that we find so captivating. Famously the males become pregnant, carrying the fertilised eggs until the fully developed young are ready to be born, at which point their new-man credentials take a turn for the worse because they play no part in parenting and simply eject the babies to fend for themselves, dumping them like heavy shopping.

Calm and composed as seahorses appear, unfortunately they have plenty to fret about. Numbers of the fifty-plus species that live in seas and oceans around the world are in freefall, with one campaign group, Save Our Seahorses, calculating that they could become extinct in the wild within thirty years. Millions upon millions are harvested every year to be used in traditional medicines, particularly in China as an alleged cure for impotence and baldness, among other things. Huge numbers are also caught for curios and souvenirs, the bony exoskeleton plates retaining their shape when dried, and for aquarium pets, regardless of their high mortality rates in captivity. International trade in seahorses has been restricted, and the buying and selling of them even banned on ebay, yet the unsustainable killing shows little sign of slowing.

Here in Britain, seahorses have been protected by law since 2008, which may come as a surprise, not because it took so long to introduce safeguards but because we have any to conserve in the first place. While one might imagine that seahorses are exclusively tropical creatures, many are found in temperate waters and two kinds live and breed around our coast: the short-snouted seahorse and the spiny seahorse. They are so rare and unobtrusive that they escaped the attention of naturalists until the early 1800s, when confirmed UK sightings were written off as accidental arrivals from further south. However, mounting evidence, stretching back to medieval carvings, has proven that both these European species are native to Britain, and more than eight hundred records have been documented to date, predominantly along the south and west coasts, and as far north as the Shetland Islands.

Noting our wildlife sightings, and letting the right people know, is the cheapest and easiest contribution any of us can make to the conservation cause, and it has made a real difference with seahorses. The British Seahorse Survey, launched by Neil in 1994, has logged finds by snorkelers, divers, fishermen and even beachcombers chancing across dead specimens, all of which have supported the case for their resident status and legal protection under the Wildlife and Countryside Act.

One of the most significant sightings came in 2004, when a pregnant male spiny seahorse was photographed amid the eelgrass of Dorset's Studland Bay. It suggested the exciting possibility that this could be a breeding site for the species, and further searches found plenty more. A number were even tagged with tiny numbers slung on loops of elastic around their necks, and up to forty were identified on summer territories dotted around the seagrass beds, with some individuals present year after year.

'You tend to have guidelines, rather than rules, for spiny seahorses, because in other areas they can behave differently,' Neil said. 'But we know that the Studland seahorses spend the winter in deeper water, safe from storms, then return when the inshore sea warms up and form pairs for the season.'

The spiny seahorse is a handsome species, longer than a biro and sporting spiky bristles on its crown and back that help it blend in with its surroundings. Its pair bonding is, apparently, quite touching to observe. Every morning the female pays her mate a visit in the small patch of seagrass where he resides, and they swim side by side, heads bowed, dorsal fins fluttering, changing colour from darker tones to a lighter shade of yellowy brown and mirroring each other's moves. If mating is on the agenda, she will deposit her eggs in his front pouch, and of the five hundred to one thousand fry born a month later it is estimated that two are likely to survive over a couple of years to reach sexual maturity. This sounds like a low success rate, but considering he can pump out several broods in a summer it's not bad going for a fish.

Unlike the smooth-skinned and stockier short-snouted seahorse, which can be found across a range of seabed habitats, the 'spiny' depends on seagrass beds for its survival and this has created something of a conflict in Studland. On the one side are the yacht owners, for whom this sheltered bay has long offered a safe and scenic place to stop over. On the other are marine conservationists, who argue that anchors and mooring chains are damaging the seabed and more needs to be done to protect the seagrass and seahorses that depend on it.

While the debate rages the seahorses have, for whatever reason, been steadily disappearing from the area. Neil told me that the peak total of forty individuals had fallen over the years to barely half a dozen. It had become increasingly difficult for the experienced team, with one thousand hours of underwater survey work at Studland between them, to find even a single spiny, so when I informed Neil that I couldn't dive he paused to sip at his tea and think things over. As he did so, I rambled on with polite suggestions and eventually he agreed to my plan: I would hire a sea kayak and accompany the crew at the surface, making myself useful by warding off any approaching motorboats, then lower myself into the water with a face mask at the ready if a seahorse was spotted.

It was midsummer when we finally met at the site, several days after my fen raft spider encounter, and I was concerned to hear that no seahorses had yet been located. Setting off from home at 5.30am, I had arrived at the National Trust car park in the village at 9am in time to meet Neil, and we changed and walked down to the empty South Beach, him in full diving gear and myself in a surfing wetsuit and clutching a snorkelling set hired from Studland Watersports, which had also kindly towed over a sea kayak from neighbouring Knoll Beach for me to use. The water was sufficiently warm, the visibility adequate, and Neil waded out after a few equipment checks, then, with a wave, disappeared beneath the surface in a cloud of bubbles. It would have been easy to lose him, even keeping an eye on the bursts of rising breaths. However, tied on a length of rope behind him he towed a safety buoy with a flag attached to signal his presence to boat users, and I followed this out towards the seagrass beds. A breeze ruffled the water and passing clouds threatened showers, but it was lovely all the same, bobbing about in my kayak taking in views of the wooded coastline, steep rock faces and sandy shore.

After about half an hour, Neil surfaced. 'How's it down there?' I asked.

'Nothing yet, though there are a few things about . . . spider crabs, a dragonet, sand eels and so on.'

'I'm only sorry I can't help in the search.'

Neil shook his head. 'To be honest, the visibility's not good enough for snorkelling, but it's reassuring anyway to know you're up there keeping watch.' And with a smile he ducked under and was gone.

I tracked his diving buoy west along the seagrass and back again, until eventually it turned towards shallower water and Neil stood up, pulled out his mouthpiece and began wading to the beach. 'Nothing today,' he sighed. 'It's a worry. We should normally have plenty of seahorses by now.' A sweep of his outstretched hand drew my attention to the numerous yachts anchored on the seagrass and arriving in the bay. On busy days, up to 350 boats visited the area, he told me, and a no-anchoring zone had been trialled and environmentally friendly moorings proposed to prevent damage to the seabed. 'Basically, not enough is being done to protect the seahorses living here,' he said.

The next time we arranged to meet, a fortnight later, I went directly to Studland Watersports, where I was able to hire a newly delivered kayak that had been fitted with a glass viewing panel in the bottom. The stress of the early start and long drive was swept away with each stroke as I paddled the mile across to South Beach – nothing is more relaxing than the sound of water slipping under a hull – and every so often I stopped, resting the paddle shaft on the rim with a hollow clunk, and peered through the window between my feet at the sandy seabed below, using my shadow to reduce reflections.

Neil was accompanied by two volunteers, John and Paul, and I escorted them until the water was up to their waists and they submerged. Although I caught the odd glimpse of flippers and air tanks through the kayak glass, I kept my distance and concentrated on scanning the bottom for signs of life.

It soon became obvious that any seahorses were likely to be concealed from my view beneath the long strands of eelgrass combed over by the currents. In addition, Neil had told me how they reduced their profile by turning away from approaching danger, and this, coupled with their slow movements and excellent camouflage, made them extremely difficult to spot, which is one of the reasons adult seahorses have few predators – not that such skinny and crunchy morsels would make for much of an appetising meal in any case.

The three divers covered a lot of ground, but once again there were no seahorses to be found, and after a quick debrief and some collective shoulder shrugging, I paddled back to Knoll Beach and drove home disappointed.

A week later I was back, only the sea had been stirred up by winds from the east and visibility was so bad I could only see a couple of feet down through the kayak's glass bottom. By the time I reached South Beach, Neil had assessed the conditions and decided it wasn't worth it, and I had to turn around and hope for settled weather another day.

It didn't come. Two further surveys were cancelled because the water was too cloudy, and I was beginning to wonder whether, in attempting to find a seahorse, I was out of my depth, quite literally. I couldn't dive, none had been spotted, and I felt a bit foolish driving such long distances in order to paddle about at the surface without even getting wet. Perhaps, I thought, it would be sensible to abandon the search and pretend it never happened, strike 'seahorse' off my list and find a replacement rarity that made life a little easier.

So, let's just forget I ever mentioned seahorses, shall we?

Then, encouraging news filtered through that a spiny seahorse had been seen off Swanage and another in Poole Harbour, and a short while later I received an exciting email from Neil telling me that two pairs had at last been located in Studland Bay. I could hardly wait to get back there. Did I say forget about seahorses? I must have been misquoted.

Only there was a snag: I was unable to make the next survey visit, so I decided to ditch the kayak, take the plunge and try snorkelling in the hope I got lucky – which is why I found myself swimming alone over the seagrass beds, with sand eels and pipefish for company, before being struck by cramp and struggling to get ashore. Challenging as it had been, it finally felt as if I was sharing in the search for seahorses and experiencing this special marine habitat in a more intimate and meaningful way. I promised myself that I would return to Studland, as I desperately wanted to see a seahorse in the wild. However, that would have to wait. I really needed to turn my attention to other scarce animals, and it was high time I paid Wales a visit.

There are plenty of unwritten rules in journalism, and one is that it is compulsory to make puns when writing about bees. Even serious articles about the decline of these vital pollinators are invariably topped with headlines incorporating a little friendly wordplay. This may be no bad thing if it means we associate bees with humour rather than their ability to sting, though it does depend on how painful the puns are. After years of repeated exposure I think I'm gradually developing an allergy to them (I know because I break out in hives), so I hereby vow to resist temptation. Instead of droning on like some bumbling idiot, I will do my best to comb all comic buzzwords from my thoughts, draw a veil over past lapses and wax lyrical without resorting to jarring honeyed prose swarming with puns.

Bees are fascinating insects. I have kept two or three colonies at the end of my garden for a number of years, and while they haven't repaid my hospitality with much in the way of produce, they have brought me a great deal of pleasure. Strange as it seems, inspecting a crowded hive on a hot day with clouds of workers humming around your head can be extremely relaxing. When the nectar is flowing they are so preoccupied with their tasks that they pay you little heed as you gently lift out the wooden frames – you can watch returning workers perform waggle dances as they pass on information about the direction of their finds, or observe the queen laying eggs in empty brood cells. The sight of such a multitude of life, labouring as one, is a joy to behold. The sound of whirring wings fills your ears, and I love the smells associated with beekeeping: burning meadow grass in the smoker, warm honeycomb and the peculiar aroma of tree resins used by the bees to seal gaps in their cedar-wood boxes.

It can be a dispiriting hobby on a wider scale, though, thin on good news and preoccupied with disease control. Determined to keep enjoying having bees, I now steer clear of clubs and apiarists' magazines, and meddle less and less with my colonies. I like to think of my old hives in the same way as bird boxes – they are there to provide nature with a home.

Honeybees live in large colonies all year round, which is why they need to stockpile such impressive stores. Pollen provides a source of protein, while sweet carbohydrate nectar is modified with enzymes and concentrated through evaporation to create long-lasting honey. Bumblebees, on the other hand, establish a small colony from scratch every spring, as only mated queens make it through the winter. No honey hoards for us to plunder. However, they perform a hugely important service as pollinators, so much so that tens of thousands of colonies are imported into Britain every year from Europe for use by commercial growers of crops including tomatoes and strawberries.

In the wider countryside, bumblebee numbers have plummeted in recent decades because of changes in agricultural practices and the loss of wild flowers across our landscapes. Two of our native species have disappeared altogether since 1940, though one is being helped back. Short-haired bumblebees have been reintroduced in Kent after the UK population was declared extinct in 2000 following more than a decade without sightings. And guess where we got the new queens from? Our friendly pool-frog suppliers: Sweden. I can just imagine the telephone conversation:

'Hi Sweden, this is Great Britain here. How's it going?'

'Hello. Again. What have you lost this time?'

'Lost? Well, okay, bit embarrassing really . . . all of our short-haired bumblebees.'

'And I suppose you want some of ours?'

'No, no, it's just . . . well, yes – but only if you've enough to spare.'

'So let me get this straight, we've given you pool frogs, not to mention large blue butterflies and a few red kites in the past, come to think of it, and now you want short-haired bumblebees?'

'Only a few, but we promise we'll be good this time. Honest.'

Just as well someone's been looking after their wildlife.

Despite widespread declines in bumblebees, we have gained a newcomer in recent years: the tree bumblebee, which made it across the Channel from mainland Europe in 2001 and has spread rapidly, typically nesting in bird boxes. It brings our total to twenty-five species, including common and widespread kinds, such as the red-tailed bumblebee and garden bumblebee, and a worrying proportion that can now be found only in small numbers. Among these scarcities is the shrill carder bee, the rarest species in England and quite probably all of Britain – and the one I hoped to find.

While most of our bumblebees are black, with their tail-ends tipped white or reddish brown and bodies marked with bands of yellow, four species are ginger-and-buff coloured and belong to a group known as carder bees. They get their name from the wool-processing term 'carding', because they comb grass and plant material together in order to make their nests low to the ground or in old mouse holes, and the shrill carder is so called for its noticeably high-pitched buzzing.

At one time, shrill carder bees could be encountered across most counties of England and Wales, and were casually referred to by one naturalist in his early-twentieth-century field notes as being 'everywhere as usual'. By the late 1990s, searches of old haunts struggled to spot any at all – little wonder given that we had wiped out ninety-seven per cent of our flower-rich grasslands in the intervening years, as farming techniques intensified and sixty-four thousand square kilometres of meadows got the chop. Today, these fairly small bumblebees are confined to a handful of locations, including Salisbury Plain, the Thames Gateway and south Wales, which is where I headed to meet Sinead Lynch, conservation officer for the Bumblebee Conservation Trust.

We had agreed to go searching for them at Newport Wetlands, a nature reserve run by Natural Resources Wales overlooking the Severn estuary, and we couldn't have picked a better day. The car park was filling up and families were out in force in the sun, flocking to the main RSPB visitor centre for a browse in the shop and a look at the friendly ducks before heading off along nature trails through the reed beds. Sinead arrived with net in hand, and I slapped on sunblock and a cap and we walked slowly up the main path into the reserve. The verges and meadow areas had been managed with bumblebees in mind, and several could be seen foraging amid the clover and brambles. Sinead pointed out small ones with reddish-tipped abdomens called early bumblebees. They worked the blooms methodically, and each would leave a faint scent as it touched down on the petals to feed, which indicated to others that the flower had just been visited and was not worth a look. By the time the flower's nectar supply had been replenished, the smell would have faded.

'They're amazing insects,' Sinead said, when I asked what drew her to bumblebees. 'They can be a bit challenging for a beginner to identify, but not so hard that you want to give up, and they have really interesting colony life cycles. And of course they're cute and furry!'

'Cute' and 'furry' are not the kinds of terms scientists normally bandy about, but they certainly count for a lot when it comes to conservation. 'People generally like bumblebees,' Sinead went on. 'They don't bother us as they go about their business, and they're also a familiar sight in people's gardens and very much the sound of summer.'

Plump and hairy, bumblebees are well adapted to our northern climate. They're up and about early on cool mornings and late to bed in the evening despite falling temperatures, and are even able to shiver to warm up by decoupling their quivering flight muscles from their wings – a bit like revving a car engine with the clutch in. It is these high-speed muscle vibrations, twanging at two hundred times per second, that make their buzzing sound, rather than the actual wingbeats.

Even though it was midsummer, Sinead had yet to see a shrill carder bee, and we spent a long time inspecting the flower patches, checking out everything that flew by. One open area, a former dumping ground for fuel ash overlooked by a nearby power station and giant pylons, didn't exactly have the most picturesque views, but the soil had given rise to an interesting mix of plant life supporting plenty of bees, and, wading into the vegetation, she found a common carder feeding on comfrey. A warm brown colour, with quite a dark abdomen, this was a female worker collecting food for a brood hidden in a nest somewhere.

The typical year for a bumblebee colony starts in the spring, when a groggy queen crawls out of her winter hiding place and feeds up before looking for a nest site, which is when she can be spotted zigzagging low to the ground prospecting for somewhere suitable. Once it is found, she creates a nest and wax pots to store food, before laying eggs and rearing her first clutch of female workers, which take over the provisioning duties – and this is what was occupying the common carder bee we were watching. Then, later in the summer, the colony raises the next generation of queens and males that head off to find mates. Only newly fertilised queens survive into winter to start the cycle again the following year.

'Shrill carders are greyer looking, and reasonably easy to identify once you've seen one before,' Sinead said as we moved on, adding that the species tended to emerge from hibernation comparatively late and that queens could still be on the wing.

As we returned to the path, she called out to the reserve manager, Kevin Dupé, who happened to be passing by on a cycle, and he stopped to help out.

'This is one of the best places in Britain to find them,' he said, 'but I've only seen two so far this year.'

It turned out that one queen had been feeding the previous day in a field away from the main trail, and he led the way, pulling aside a section of fence and guiding us down through high grasses to an area of tufted vetch and grass vetchling.

We stood around chatting in the sun until I noticed a bumblebee arrive and settle on the vetch.

'There's something,' I said, with no idea what it might be.

Sinead spun around. 'Ah,' she said, quickly lowering her net over the purple flowers, 'that's one.' She held the end up so the bee flew deep into the mesh bag then transferred her catch to a container, and with a shared sense of relief and delight we were able to admire the find: a queen shrill carder bee. It was indeed greyish looking at arm's length, perhaps more straw-coloured close up, and a dark band ran between its wing bases while the hairs at the tip of its tail were orange. Very different to any bumblebee I had seen before.

'It is always great to see one, and especially a queen,' Sinead said. 'It means they're still clinging on here, and they're lovely looking.'

We let the bee go and it flew off the way it had come.

After waiting to see if any others were about, we headed up to the main path. A quick circuit of the reserve drew a blank; however, a return visit to the field found a queen feeding within a foot or two of the original spot. Given that it departed in exactly the same direction, presumably to a nest a short way away, one could assume it was the same hard-working individual, toiling to raise its brood, all the while keeping its kind from extinction in Britain with the help of an unassuming area of bee-friendly flowers tucked away in a safe and secluded corner of a reserve. It shows how much we have lost that a small meadow can make such a big difference.

I, too, would be returning to the same patch, or near enough, within a week, although in slightly less hospitable conditions. The excellent bumblebee sighting meant I had seen my five target rare and endangered invertebrates: the Duke of Burgundy butterfly, streaked bombardier beetle, Norfolk hawker dragonfly, fen raft spider and shrill carder bee. I could have put my feet up and toasted a job well done. However, there are so many more scarce creepy-crawlies to choose from that I couldn't resist adding just one more insect.

It was the evening of 29 July 1972, and Welsh entomologist Dr Neil Horton had driven the twenty or so miles from his home to a disused upland quarry in a secluded valley north of Abertillery in Blaenau Gwent. The site overlooked a farm set among fields of lush pasture, and on either side the ground rose steeply beyond stone wall boundaries and a scattering of trees to exposed moorland above. It was an attractive, out-of-the-way place and he was looking forward to the night ahead and the chance to shed a little light, quite literally, on the area's wildlife secrets. However, the weather was breezier than expected and another vehicle was already parked next to the single-track road, in the exact spot where he planned to place his trap. The occupants may have been tourists or lovers seeking a little privacy away from prying eyes. Either way, his knock on their car window gave them such a fright that they drove off, leaving him to set up his equipment beside the narrow gulley that ran up from the overgrown quarry to high ground. His trap consisted of a wooden box with a hole in the top, above which a battery-powered mercury vapour bulb was positioned between baffles. Anything flying into the light would be funnelled into the container beneath, where a number of old egg cartons provided hiding places. As it began to get dark, he switched on the lamp, illuminating a wide circle of surrounding bracken, bilberry and heather, and waited for fluttering life to appear out of the blackness.

Aged in his late fifties, Dr Horton had been fascinated by moths and butterflies ever since he was a boy growing up in Monmouthshire. He had left the area to study medicine at Cambridge, and later botany and zoology for good measure, and served as a doctor in the RAF overseas before returning to Wales, where he was able to further his passion for entomology, accumulating lengthy lists of local finds and eventually becoming Monmouthshire county recorder for Lepidoptera. On this evening, eager to survey moorland moths, he had come to the Tillery valley to see what resident moth species were about, choosing the disused quarry site because it was out of the wind, and hardly expecting any great surprises.

As moths appeared, attracted to the light, he began noting the various kinds, as he had done diligently down the decades, until one turned up that he simply did not recognise. It was not particularly striking – reddish brown and thick-bodied, with a wingspan of about three centimetres, each forewing bearing a light, kidney-shaped mark and a couple of wavy lines – yet Dr Horton was unable to put a name to it. For someone as experienced as him this was unusual.

Stumped by the mystery moth, Dr Horton turned to experts at the Natural History Museum for assistance, and they identified it as a species resident in upland areas of Europe. Not Wales. In fact, it had never been seen before in Britain, and as such had no common name, just a scientific title. The honour fell to Dr Horton to christen this significant first for the country, and he called it the Silurian moth, after the warlike Silures tribe once found in this part of south-east Wales.

No more were found for several years, which raised the possibility that it was a one-off vagrant, until Dr Horton eventually came across the species again in the same upland valley, confirming that this new discovery was indeed a British resident that had somehow evaded the attention of naturalists over time, probably due to its plain looks and the remote location. Dedicated moth devotees were keen to see this Welsh specialty for themselves, and over the years that followed, regular records trickled in. But its scarcity and restricted range were a cause for concern. A moorland fire, or new grazing regimes altering hillside plant life, had the potential to wipe out all of our Silurian moths. Localised rare animals are extremely vulnerable to extinction. Habitat changes, harvesting, predation and outbreaks of disease or parasites can all prove devastating, while small populations also run the risk of inbreeding – it is the same for shrill carder bees and seahorses, and for common skates and smooth snakes. The hope was that Silurians were more widely spread across the uplands in this part of Wales, and perhaps beyond.

Thirty-three years after Dr Horton's initial discovery, and four months before his death at the age of 89, caterpillars were found for the first time feeding on bedstraw and bilberry high up and late at night in icy spring conditions. And in 2011, a summer light-trapping expedition further north in the Black Mountains of Brecon Beacons National Park located flying males along the elevated Hatterrall Ridge on the border between Wales and England – proof at last that the species was not just confined to the original Gwent valley.

The Silurian moth is a species of tough terrain and long nights, which tests the endurance of the hardiest nature enthusiast, and I was looking forward to the challenge of accompanying moth expert Dave Grundy as he carried out further survey work along this ridge. However, he warned me in advance that I should be prepared for a climb, cold weather and to be out until dawn, because the moths tend to fly well after midnight and it was safer to return along potentially treacherous paths in daylight.

We had arranged to meet on a Friday, and I hurried to finish work promptly in order to get going on the long drive.

'Off anywhere nice?' I was asked by a colleague as others shared their weekend plans.

'Just a night out, somewhere quiet,' I replied, truthfully, and spared the details, dashing off to my car and racing the sun as I drove north to Bristol, over the Severn Bridge and on past Abergavenny along winding roads to Longtown, where, true to form, I took a few wrong turns and got lost, wasting half an hour. Finally, I pulled up at dusk in a parking area beneath an escarpment called Black Darren, changed into my outdoor clothes and set off as quickly as I could before the evening light faded completely.

It was a tough climb along a steep diagonal path to the top of Hatterrall Ridge, but worth it for wonderful views west over the Vale of Ewyas and east across Herefordshire and the upper reaches of the Olchon Valley, where square fields bordered by trees looked to have slipped down the near-vertical hillsides and come to rest at the bottom. The fresh air, scenery and grass beneath my feet recharged me after a long working day, like green energy for the soul, and I was thankful that a little brown moth had provided the excuse to pay a visit.

Offa's Dyke runs along the middle of the narrow ten-mile ridge, and I followed a deserted path that traced the medieval boundary, checking my map and written directions in the gathering gloom and looking for a 610-metre triangulation point. Up ahead, scattered bright lights were burning in the twilight, and after passing the concrete pillar I carried on towards them, drawn like a moth, until I arrived at a trailer by the path, with a generator motoring away beside it and cables running out to several lamp traps dotted around in the heather and cotton grass. It was a bizarre sight in the middle of nowhere, like a runway lit up ready for the arrival of an alien spacecraft, and Dave Grundy was nowhere to be seen. It was also freezing, so I pulled on a woolly hat and sheltered from the damp wind behind the vehicle for a while before flicking on my head torch and deciding to check he hadn't been abducted. There were other lights further off, and I kept to the path and walked on into the darkness until I saw someone coming towards me, illuminated from behind with hood pulled up.

'Dave Gundy, I presume,' I said, shaking his hand.

Just as well it was. It certainly felt the strangest of meetings, and I didn't get to see what he looked like – a friendly face with beard and glasses – until we arrived back at the trailer and chatted next to one of the moth-trap lights.

'You have to be a bit mad to be out here doing this,' he laughed. 'It's what I like to call "extreme mothing".'

Dave was among the pioneering team that discovered Silurians on the Hatterrall Ridge, and his first sighting was a stroke of luck. 'I was on the ground, slithering down to a trap on a steep section below the ridge, when I saw one next to it – not even inside. It could easily have been missed.'

He was uncertain about our chances of success this night, however, because the long, cold spring, a distant memory now, was likely to have slowed the moths' life cycle and delayed the emergence of flying adults. Added to that, the ten traps, running at fifty-metre intervals, were covering ground that had not been surveyed before, and the breezy conditions were far from ideal. There was nothing to do but hope for the best, sit in the trailer out of the wind and kill time talking about moths, as you do.

Dave, who worked in conservation before launching his own wildlife consultancy, was originally a keen birdwatcher until his interests diversified. 'One of the great things about moths is that there is so much to see and learn,' he said. 'You can actually make a contribution to scientific understanding, which is what makes it so exciting – every time you open up a trap you have the chance of finding something of real significance.'

Living in a terrace in Birmingham with a concrete backyard, he relishes the opportunity to explore the countryside as part of his nature-surveying work, and has grown used to sleeping by day and being alone outdoors in the dark. I admired his nerve. It's easy to tell yourself there is nothing to be afraid of at night, but who can be sure, what with all the moth catchers, bat- and great-crested-newt surveyors, corncrake counters, natterjack-toad recorders and pine-marten spotters roaming the countryside.

It was after midnight when we set off to check on the traps, opening the lids, reaching into the drums and lifting out the cardboard egg trays like a game of lucky dip – only neither of us claimed the main prize. 'Disappointing,' was how Dave summed things up. No Silurian moths, and just a handful of other species. But they were enough for a relative beginner like me to be getting on with. While I can recognise a few of the more familiar species, I was struggling to identify any of those caught, and worked hard to memorise the catches Dave showed me: narrow-winged pug, glaucous shears, map-winged swift, red twin-spot carpet, light knot grass, iron prominent.

The assorted names of moths are one of the most charming aspects of our natural history. Taking inspiration from places, food plants, markings and the whims and fancies of Victorian entomologists, they have a sense of stories half told, a nostalgic quality that connects you with the poetic imaginations of the past. True lover's knot, powdered quaker, Setaceous Hebrew character, dingy footman, toadflax brocade, maiden's blush, feathered gothic, scarce merveille du jour, frosted green, smoky wainscot . . . you want to say them as much as see them. And then there are the amusing ones: the suspected moth, the anomalous, the uncertain and even the confused.

Such variety of names reflects the sheer diversity of our moth species. People who think they are universally grey, brown and boring are well wide of the mark. Okay, a few hundred species may, admittedly, be a little on the dull side. But given that we have around 2,500 in Britain to choose from, including nine hundred of the larger so-called macro moths, that leaves a staggering array of attractively patterned specimens to marvel at. Some, such as the garden tiger moth, cinnabar and elephant hawk-moth, make many butterflies look decidedly ordinary. Others are masters of camouflage, such as the buff-tip moth, which resembles a twig, and the Chinese character moth that could be easily mistaken for a bird dropping. And then there are myriad kinds with subtle and exquisite markings that look as if their wings have been finely dusted with colour, pressed against lacework or embroidered with threads of precious metal.

'One of the great things about them is that they allow you to experience nature up close,' Dave said as he showed me the catches. 'Unlike so many other wild animals, you can actually hold moths and have beautiful and impressive specimens sitting contentedly on your hand.'

Moths are perhaps an acquired taste, like anchovies and organ music, and largely misunderstood. For a start, very few species eat clothes. We have a tiny number with an appetite for fashion, and they have declined because synthetic fabrics are not to their liking. Moths are also not exclusively creatures of the night. In fact, there are more day-flying moths in the UK than there are butterflies, and some butterflies fly after dark – the red admiral among them. Differences between the two groups are actually less clear-cut than one might imagine. Butterflies have club-shaped antennae and mostly hold their wings together above them, although resting skippers look every bit like moths, while day-flying burnets have butterfly-like antennae.

After a couple more hours chatting in the trailer, we braved the elements for another light-trap foray, and as the sun came up a final round of checks forced us to admit defeat. No Silurian moths. For now at least. Dave planned to return the following evening with a few others in order to survey a more southerly section of the ridge, after the national park authority had helpfully towed the trailer-load of traps and generators to a new spot, and he was happy for me to join him.

Once packed up and off the hill I drove to a quiet lane and managed a couple of hours sleep in the car before lunch, then in the evening headed for the agreed meeting place, the remarkable Skirrid Mountain Inn in the village of Llanfihangel Crucorney. Dating back to the 1100s and reputedly Wales's oldest pub, it once acted as a courthouse where more than 180 criminals were put to death – many hung in the stairwell that still bears rope scorch marks in the old beams. Little wonder the building is said to be haunted. Pushing open the centuries-old front door and passing a disturbing model figure of a magistrate in the hall, I had a good snoop around, climbing the warped staircase where the steps, held in place by wooden pegs, creaked and groaned with dreadful memories. The shadow of a judge is said to patrol the landings, while the smell of lavender betrays the ghostly presence of a resident who died in one of the bedrooms long ago. This was not your average pub.

After an early meal downstairs I waited for the group to turn up, playing a private game of 'spot the moth-catcher' as people came in through the bar door. My guesses were hopelessly wrong. Then Dave arrived with six others and we drove in convoy to a parking spot at the tail end of Hatterrall Ridge and hiked to the top. The group included a retired bus driver, a teacher, a student and a keen photographer, and they came from places such as Wirral, Shropshire and Swansea, all hoping to see a Silurian moth.

It was warmer than the previous night, though there was still a little wind dragging lumps of black cloud overhead. The traps were spread out across areas of scree and nestled in among the thick bilberry along the sheltered eastern edge, and as night fell the first few moths began spiralling in. It is believed that moths confuse a light with the moon, which they use to navigate. While they can keep the moon in the same position relative to their flight path, they are forced to adjust their direction continually to do the same when they pass a lamp, flying in ever-decreasing circles until they end up colliding with the bulb, dazed and confused.

I followed Dave around as he checked the traps, and it seemed we were in for a good night. Numerous moths were among the egg trays or on the outside of the drums, and I was amazed at his identification skills as he reeled off the various names. It was also surprising how many species were up here on this high moorland ridge, not only zipping through the pools of light but also caught in our torch beams as we wandered around.

Over the last century, Britain has lost more native species of moth than bird, mammal, amphibian, freshwater fish, dragonfly, bumblebee and butterfly extinctions added together. Few people care. We tend to think of moths as suicidal in any case, throwing themselves at windows and hurtling headlong into car headlights, but overall population declines are telling, even given new species colonising from overseas. Moths play an important role as pollinators and as part of the food chain, with a pair of blue tits, for example, needing a whopping fifteen thousand caterpillars to raise a single brood of chicks. So to witness such diversity in what seemed a bleak and inhospitable place was heartening. Not many people would choose to spend a Saturday night up in the Black Mountains looking for moths – eight of us to be precise – but it was well worth the effort and I could hardly keep up with the array of new finds Dave and the others kindly showed me: eyed hawk-moth, lobster moth, puss moth, peppered moth, lesser swallow prominent, beautiful brocade, grey mountain carpet . . . By the time the sun came up and the red grouse started calling, the team had recorded three hundred moths from seventy different species, with a few notable scarcities among them – although, unfortunately, no Silurian moth. The ultimate mountain rarity had eluded us. Still, I had thoroughly enjoyed looking, even if I was so tired after forfeiting two nights' sleep that I could hardly remember my own name.

Foiled by seahorses and Silurians, and with some extremely tricky animals still remaining on my list, I needed a boost – and I took a chance that paid off. I drove all the way from Devon to Kent and successfully tracked down one of Britain's rarest and most monstrous insects: the wart-biter cricket. As with the Silurian moth, it was an added extra on my invertebrates list, a bit of a spur-of-the-moment decision. Failure would have left the feeling that luck had turned against me, however ten minutes of unbroken sunshine ensured good fortune became a friend once more.

Obviously it sounds like madness to drive three hundred miles just to see a chirping cricket, but then I couldn't make any great claims to sanity, having spent two nights on a hilltop looking for a moth. In any case, the wart-biter is no ordinary cricket. Found in fewer than half a dozen sites in southern England, this large and fearsome-looking rarity not only eats plants, but is also big enough to dine out on other crickets and grasshoppers, and its carnivorous tendencies and powerful mandibles were once employed in Scandinavia to chew off warts – hence its name. The pioneering Swedish taxonomist Carl Linnaeus dubbed it the 'verruca devourer', with the specific epithet *verrucivorus*, which sounds like an alarming kind of dinosaur. However, experiments have found its crude and painful surgery to be ineffective – and in a sponsorship masterstroke the UK's top-brand wart cream company Bazuka has come to the aid of this endangered 'competitor', funding habitat-restoration work by the Species Recovery Trust to prevent its extinction.

After the long drive east I arrived late at night in the village of Temple Ewell, near Dover, slept in my car and the following morning made my way to the nearby Lydden Temple Ewell National Nature Reserve, an area of ancient chalk grassland managed by Kent Wildlife Trust on the south-facing side of the Dour Valley. The fields were filled with flowers and a profusion of blue butterflies, and while I waited for reserve warden Pete Forrest to arrive I explored the mown pathways – though at a fairly slow pace given I was limping.

A few days earlier I had been chased by Highland cattle while running with my dog on Dartmoor and come a cropper. Despite their intimidating horns, Highland cattle are generally laid-back creatures that scarcely break into a fast walk, let alone a run, so I was surprised at their turn of speed when I foolishly attempted to take a shortcut between them and their calves thirty yards away. Half a tonne of beef with a low IQ and strong maternal instinct is not to be taken lightly, and three of them came charging down the slope at me, at which point I turned, slipped and fell, gashing my knee on a rock. My dog tugged free of his lead and squeezed under a gate as one went for him, while I clambered to my feet, dodged the horns bearing down on me and managed to jump over a cattle grid to safety. It was a fairly terrifying experience and I needed to get my knee checked over at the local minor-injuries unit. The senior nurse examined the joint, then asked me to rate the pain on a scale of one to ten. Not wanting to exaggerate, I gave it a four, which sounded like I didn't really need to be there, and she bandaged it up and sent me on my way.

Wart-biter numbers at Lydden Temple Ewell had also been sent tumbling by livestock, though in a more roundabout kind of way. At one time, pigs were kept on a field nearby, and their feed attracted crows and magpies that loafed around during the day in trees on the reserve. These perches not only overlooked the pig pens, but also happened to be right next to grassy areas where the wart-biters lived, and the heavyweight invertebrates were gradually picked off until, by the early 1980s, none were left. It took the closure of the piggery, and dispersal of its attendant crows, before the reserve could be considered safe once more for the crickets, and under a Natural England Species Recovery Programme, replacement wart-biters from a Sussex site were reared and released by experts from the Zoological Society of London. Two decades on, the protected population is still going strong.

At the agreed time I found Pete, who was busy organising conservation tasks for a group of volunteers, and once he had set them to work we visited the section of the reserve where wart-biters were known to live.

'They've been heard here recently, but they can be very elusive,' he said, admitting that despite working on maintaining habitat for wart-biters, he had never actually seen one before.

For the next hour or so we searched the tussocky grass around low clumps of blackthorn, my injured knee making it difficult to bend down, and we came across several species, including a Roesel's bush-cricket, an impressive dark bush-cricket and a common green grasshopper. Crickets can be told apart from grasshoppers because they have long, thin antennae and 'sing' by rubbing their wings together, which a few of them were doing, while grasshoppers have short antennae and rub their hind legs against their wings to make their sounds. This distinctive stridulation, as it is known, is a useful guide to identification, and the wart-biter has a loud and repetitive chirp much like the clicking of a freewheeling bicycle. Pete played me a recording so that I knew what to listen out for, and he had also come equipped with a bat detector to help make the shrillest noises audible. He lent it to me, tuned at a frequency of 20kHz, while he went off to check on the volunteers. You know you're getting older when you need the assistance of a device set to the level of a typical child's hearing in order to make out higher-pitched sounds.

Wart-biters call only when it is hot and sunny, and cloud cover meant it simply wasn't warm enough for them. After hobbling around the field with my giant handheld hearing aid for a while, I began to feel that perhaps I was wasting my time. Then I got lucky. A break in the clouds let the sun beam down from a blue sky and the temperature rose rapidly. It was enough to make the difference. I picked up a repetitive chirping on the bat detector, every bit like the wart-biter sound Pete had played me, and as I walked towards an area of blackthorn and brambles it got louder. I turned down the detector volume and was relieved that I hadn't really needed it after all, as the chirping was clearly audible coming from a patch of vegetation a few feet away. Treading softly, I scrutinised the ground until I noticed something creep slowly out from a shadow into the open. Leaf-green and far bulkier than the other crickets I had seen, it was a robust and handsome beast with spotted sides, a front end that looked almost armour-plated, and long legs bent double ready to leap away from any danger. Unmistakably a wart-biter. And I could actually see its wings rubbing together as it sang, until it noticed me and immediately fell silent.

I called out to one of the volunteers at the top of the field: 'Quick, get Pete! Tell him I think I've found something!'

They headed off, and a minute later Pete, accompanied by a fellow conservationist, came jogging down towards me and I carefully pointed out my find.

'Yes, that's a wart-biter!' Pete beamed. 'Brilliant!' And we all shook hands, grinning as we celebrated our first-ever sighting, while the star of the show quietly sidled off into the undergrowth.

I stayed on in the field long enough for another break in the clouds to occur, and as the sun nudged up the heat it was like a switch had been flicked. Within seconds I could hear not one, but two male wart-biters singing, and watched as one crawled up a sprig of hawthorn into full view, cleaning his mandibles while his wings scratched out his summer song. Definitely worth a day out in Kent, and I left for home feeling, well, pretty chirpy.

# Eleven

The lengthy kit list was like none I had ever read before. In addition to the kind of essentials one would normally expect to pack for a trip to the Hebrides, such as waterproofs, walking boots and warm clothes, it included 'goggles', 'hi-vis fluorescent waistcoat', 'knee pads', 'gardening gloves', 'spring balance', 'number 1 and 2 pliers' and 'boiler suit'.

Fortunately I wasn't required to bring all of the more unusual items, which was probably just as well. The last time I had flown to Scotland, airport security staff confiscated my cartons of juice and yoghurt as a potential terrorist threat, so I'm not sure what they would have made of pliers, goggles and a boiler suit tucked in my luggage between the towel and spare socks. However, I did add something of my own to the packing that might have appeared curious in view of the fact that I was spending only a few days away: peanut butter. Three jars of the stuff.

I was travelling to the Shiants, a remote group of islands off the north-west coast of Scotland. Of all my destinations, this uninhabited archipelago set in the tidal strait known as the Minch was by far the most complicated to reach. The options for getting there and, crucially, finding a way off again were limited. Sailing myself or chartering a boat at great expense were out of the question, and a weekly day-return trip from Harris ten miles away would provide insufficient time ashore when I required several days – and especially nights – on the isles to stand any chance of tracking down my next rarity. Somehow I needed to hitch a lift, and after numerous phone calls and emails a workable solution took shape. The plan was to fly to Glasgow and then to Stornoway on the island of Lewis in the Outer Hebrides, before heading south by taxi to the Harris village of Tarbert. From there I would board the day-trip vessel to the Shiants for the outward-bound leg and remain on the islands alongside a team undertaking seabird monitoring work, returning with them on their chartered boat the following weekend. Each stage of the trip was duly booked, a cheque for my share of food and fuel was posted to the bird-ringing group, which had been responsible for the intriguing kit list, and I congratulated myself on a faultless feat of organisation. What could possibly go wrong?

Every day I checked the weather forecast, and every day it got worse and worse as the date for departure neared and storms moved in from the Atlantic. The flights were running as scheduled – confirmed by Stornoway airport, where, for some reason, customer service assumed I was a pilot calling to ask about conditions and transferred me to an air-traffic controller, who started babbling on about approach velocities until I explained that I was a passenger. It was the boat trip I was most anxious about. The window of opportunity to motor across the Minch was narrowing as angry masses of tight isobars shouldered their way east, sandwiching skinny spells of calm weather and threatening to rule out travel altogether. Getting there would be like squeezing between bullies in a corridor – there was little room for manoeuvre.

Disaster then struck when I finally arrived at Glasgow airport. Switching on my mobile phone after the flight, I discovered a text message from the boat owner informing me that because of the forecast he needed to use precious gale-free time the next day to voyage to St Kilda instead. I was stuck. There was no point in continuing on to Lewis. It seemed that the Shiant Isles, a place I had heard so much about and had been looking forward to visiting for weeks, were destined to remain out of reach and I would have to make for home.

There was, however, one slim possibility. A BBC documentary crew were travelling to the islands at the same time to film puffins for the series *Coast*. I knew because they had contacted me earlier, as they were interested in my reasons for going, so I gave one of the contacts a ring. Delayed by the rough conditions, it turned out that they were due to cross the following morning, and they were happy to offer me a ride over on their chartered boat if I could get to Uig on the Isle of Skye by 9am. Easier said than done. I needed to start running – not the 225 miles north to Skye, but in order to catch a bus. The six-hour service was about to leave and it meant a sprint from the booking office to the coach stop, clutching a ticket for the last remaining seat and a guide to local accommodation, in order to make it aboard with less than a minute to spare. It took most of the journey before my heart rate returned to normal. After a night in Portree, a disturbed one because the hotel fire alarm went off at 4am, I covered the final few miles by taxi to Uig harbour, where I was relieved to find the BBC crew – presenter Miranda Krestovnikoff, cameraman Dan, researcher Phyllida and soundman Brian – and their waiting skipper Andi on his rigid inflatable boat piled high with filming equipment.

'Have you got the peanut butter?' Miranda asked with a smile.

'More than enough,' I laughed.

The weather was fine and the Shiant Isles awaited. I was off to find a rat.

A rat? Why on earth would I come so far in order to see something as widespread as a rat? Well, there are two kinds in Britain: the abundant brown rat and the black rat, which, incredible as it sounds, is now one of our rarest mammals. For a pest species of such notoriety, that may be hard to believe, especially as they are extremely common elsewhere, yet it is true. Numbers of British black rats have plummeted from the multimillions to a few thousand, and while they occasionally turn up in dockland areas, most likely arriving on incoming cargo ships, the Shiant Isles are recognised as having probably the last stable population of these creatures in the UK.

The resourceful black rat was the first of the two species to reach Britain, spreading from Asia along trade routes and arriving with the Romans. A fast learner and agile climber, it was perfectly adapted to live alongside us, scrambling up wooden beams, nesting in roof spaces and stealing scraps from beneath our noses. And the ability of the female to produce dozens of young per year, each becoming sexually mature in around three months, meant that this invader proliferated across our land.

Not only did black rats become a serious agricultural and domestic pest, eating and spoiling food stores and damaging property with their gnawing, but they also spread deadly diseases. Their fleas are blamed for the Black Death in the Middle Ages, which wiped out half of Europe's population, as well as subsequent plague outbreaks, such as the Great

Plague that swept through London in the late 1600s. No wild animal has had such a long-term and profound impact on our society or been the cause of such untold loss and misery, and it seemed inconceivable that anything could rid us of them.

Enter the brown rat. In the late 1700s, this bigger, bolder species, more inclined to burrow than to climb, followed its cousin from Asia into Britain and began to gain ground. Brick buildings and sewers suited its tunnelling, damp-tolerant lifestyle, and its aggressive nature meant that it gradually pushed out the slimmer black rat, much as the introduced grey squirrel has outcompeted the smaller red squirrel. One rodent replaced the other, while improved sanitation and food storage reduced the risk of epidemics, and our sinister plague carrier of the past was as good as gone.

Over a century ago – no one is certain of the exact date – a number of black rats arrived on the Shiant Isles, presumably stowaways fleeing a sinking ship, and they have lived on the islands ever since. There is no shortage of nutrition for them in the summer months, because the cliffs and turf slopes are home to large breeding colonies of puffins, razorbills, shags, gulls and guillemots, which provide an ample supply of eggs, chicks and scraps. But winter is another matter, when birds desert the land and the starving rodent population is kept in check by the struggle for survival. Such an annual boom-and-bust cycle has created a natural equilibrium of sorts and lessened their impact on the tens of thousands of breeding seabirds that originally were able to prosper in this lonely location due to its absence of ground predators. Accurate comparisons over time are impossible, because no one counted the local birdlife in the 1800s before the rodents came to the Shiants – at which time, it has to be said, humans were eating the puffins instead. However, some anomalies are apparent: comparatively few smaller birds, such as skylarks and pipits, reside on the isles, and breeding shearwaters and petrels are notably absent.

I was aware that time was running out for the Shiants rats. The eradication of colonies overrunning other islands in the UK, such as Lundy, and overseas has boosted populations of threatened seabirds, and such conservation measures were being talked about for this Hebridean hideaway. Black rats may have lived in Britain for a couple of thousand years – as long, or longer, than rabbits, pheasants and fallow deer – but, ultimately, they are non-native invaders and in the final analysis our seabirds will always come first. Of the two thousand-plus foreign plants and animals currently established in this country, the majority of which are considered harmless, rats have rightly earned a place on our blacklist of troublesome trespassers alongside such species as mink, Japanese knotweed and signal crayfish. Good riddance to the long-tailed vermin, many might say. Then again, *Rattus rattus* is a mammal of such profound historical significance in Britain, even if for all the wrong reasons, that one could argue it deserves to remain a part of our fauna, somewhere – though perhaps not a natural refuge as significant as the Shiant Isles. Either way, I was excited by the prospect of seeing such an infamous creature, possibly before it was too late, and I knew it wasn't going to be easy. Reaching the Shiants was only half the challenge.

The trip over from Skye was magical. Everywhere we looked there was something to see, and like tourists in a biodiversity theme park we turned one way and then the other to keep up: two common dolphins breaking the surface; kittiwakes flying low over the water; arctic skuas patrolling above. Stopping for the BBC crew to film along the way, we circled outcrops of rock crowded with bickering colonies of nesting guillemots and razorbills, fulmars, shags and gulls. Yelping, gabbling, grunting, cackling throngs amid the boulders. And we spotted our first puffin – smart and comical-looking as it waddled across a grassy ledge. It is a bird of such irresistible charm that I couldn't help but smile. There were two more ahead of the boat, which dived and could be seen paddling away underwater, leaving trails of bubbles behind them, and a further couple flew past, wings whirring. This was what the *Coast* crew had come for, and as the boat headed north towards the Shiants and puffin numbers increased, it reminded me of times I had been to a football match. First you spot one or two fans in their team colours as you start along the road towards the ground; a few more here and there add to the sense of anticipation. Then scattered groups of supporters begin to merge, dozens become hundreds, until you finally enter the stadium to be greeted by a roaring crowd tens of thousands strong. Welcome to the Shiants: towering islands teeming with seabirds, a frenzied, pungent, cacophonous celebration of life.

The isles consist of three main landmasses, none more than a mile or so long: Garbh Eilean (Rough Island) and Eilean an Taighe (House Island), which are linked by a narrow natural causeway of stones, as if pinched in at the waist, and neighbouring Eilean Mhuire (Mary Island). Outcrops known as the Galtachans also run out to the west, resembling a row of stone molars. What makes the isolated islands so special are the immense cliffs of volcanic dolerite that soar hundreds of feet above the sea along their eastern flanks. Corrugated with columns created by the slow cooling and shrinking of magma, the seamed and splintered rock faces stand exposed to the elements, and debris lies at their feet: fractured pillars piled high like the ruins of shattered temples. Seabirds nest on the crags and amid the fallen boulders, or tunnel into turf that covers the treeless slopes, and such impressive land formations animated by so many birds make for a breathtaking spectacle.

Far from being exclusively a haven for nature, the remains of simple buildings and half-buried walls tell the story of human settlement on the Shiants. They have supported a surprising number of hardy souls over the millennia, who have grazed livestock on the islands' grassy backs, dug lazy beds in the peaty soil for crops and subsisted on shellfish and seabirds. The last inhabitants left in 1901, and a restored bothy on a level area of Eilean an Taighe, near the causeway, is the only dwelling.

Our boat had been spotted, and members of the bird-monitoring team that we were joining – the Shiants Auk Ringing Group – gathered on the causeway to greet us, helping ferry luggage up steps cut into the rocks to their base at the bothy. Auks are a family of seabirds that includes puffins, guillemots and razorbills, and ringing on the Shiants first began in the 1970s, shedding light on the species' movements and longevity. After a lull, efforts were resumed by conservationist Jim Lennon in 2008, with guidance from founding member David Steventon and the blessing of the Nicolson family, owners of the isles, and every summer ten or so dedicated volunteers make the long pilgrimage to continue the work over a period of a fortnight. An astounding tally of over 45,000 birds has been caught and logged so far, many having returned year after year. In 2012, a puffin bearing a ring from one of the original expeditions was recaptured and found to be almost thirty-seven years old – the oldest ever recorded in Britain.

The bothy was a low, white building with a tin roof and two windows looking out over a stony cove. Washing-up bowls, weighed down by pebbles to prevent them being blown away, sat on tables outside, along with buckets of fresh water collected from a shallow well, and I was told that a buoy hanging over a seaweed-covered rock opposite indicated the 'toilet' behind, flushed by the tide, was engaged. Tents had been pitched around the dwelling, which was being used for dining and storage, but a shortage due to flooding meant that I had to sleep inside, once a bed had been cleared of ringing equipment. As I dumped my rucksack, I noticed a sign pinned up on the wood-panelled interior, which read: 'Keep rats out. Close and bolt the front door.' Apparently they had become a nuisance around the building, and I wondered whether I would have company in the night. Shiants black rats don't carry the plague, but I hadn't been planning on getting too close.

I had time to thumb through the bothy visitors' book before we all set off for the afternoon to one of the puffin colonies. Members of the royal family had paid a brief visit three years earlier, with Princesses Beatrice and Eugenie praising the 'beautiful islands' and 'incredible views', and other entries ranged from the straightforward ('I am happy') to the perplexing ('I could get used to being uninhabited'). One youngster complained in loopy handwriting that it was 'too rainy it hearts your cheak', while author Adam Nicolson, who was the second generation of his family to own the isles before passing them to his son Tom, despaired after the rats had found a way in: 'House made a mess by the little horrors'.

Boarding the boat again, we motored around to the northern end of Garbh Eilean, gliding between watching seals to the shore. The density of seabirds was astonishing. Flotillas of auks were gathered in the glittering bay and, above us, a great wheeling mass of flying birds turned in front of the nesting areas. Every boulder had a razorbill, guillemot or puffin perched on top, all the way up to the cliffs set back from the sea, and the grassy slopes were dotted with puffins standing outside their nest burrows, some carrying freshly caught sand eels. It was here, on the steep incline, that mist nets had been set up, and after being introduced by Jim and David to the rest of the ringers – Ian, Alister, Bob, Charlie, Ruth, Alice, Karen and Carole – I sat on a tussock and watched as they worked and the BBC crew started their filming. The puffins paid us little heed, and it was possible to get remarkably close. With neat black plumage and white chests, they stood upright by their burrow entrances as if waiting in best bib and tucker for taxis to whisk them off to some formal occasion, their multicoloured beaks and orange feet adding a touch of the absurd along the lines of a spinning bow tie or tartan socks. They may have looked endearing and eccentric, but sharp claws and a powerful bill made them tricky to handle, and some of the ringers' hands bore cuts and bruises where they had been pinched and scratched. Others wore tight-fitting gardening gloves, and, donning my new pair, I was able to hold and release one of the newly ringed birds, which flew off down the slope at speed and joined the whirling flock. I also realised why a boiler suit was on the kit list when I got strafed by 'white rain' as puffins came in overhead. It is supposed to be good luck to be hit by bird shit – until it happens again, and again.

The team worked fast and efficiently, extracting puffins from the nets, squeezing metal leg rings into place with special pliers and recording identification details, and when we eventually returned to the bothy for supper, the trip's rolling bird tally written on a chalk board above the fireplace was increased to 1,765. Along with auks, there were other species listed, including young oystercatchers, sandpipers, various gulls and storm petrels. Doze off among these enthusiastic ringers and you could wake up to find a manacle-sized steel band had been fitted around your ankle.

Short nights take a long time coming this far north in summer, so there was plenty of daylight left to get organised and set about trying to encounter a rat. Nocturnal animals are never an easy proposition, but darkness lasts only three to four hours in the Hebrides at this time of year, and Shiants rats could, apparently, be seen throughout the day, in particular at dusk and dawn. The causeway was supposed to be a good place for them and, knowing why I had come, group member Alister thoughtfully placed some fish remains down near the stone steps to help entice any out.

The wind had picked up, light rain moved through and the temperature dropped as I settled in on the beach. Despite the cold, the discomfort of the wet stones beneath me and exhaustion after all the travelling, it was wonderful to find myself here. There is something immensely invigorating about remote and unspoilt places that appear so totally indifferent to human existence. From the achingly slow processes that created these islands and beat them into shape, to the frenetic seasonal invasions of breeding birds, the cycles of creation and destruction, birth and death, spinning at differing speeds like assorted cogs in the same clock, turned here with scarcely a pause for mankind. The Shiants couldn't care less about me, kneeling on a beach as if in prayer. If I died, my bones would be simply shrugged off into the sea to join all the others – toothpicks for the boulders. Yet for all that, I felt more alive and significant than ever. The middle of nowhere is as much the middle of everywhere, and with yourself at the centre you can imagine you're the last person on earth, facing the gods alone. They might even notice you here.

I had found a spot low down on the causeway, which was sheltered from the worst of the wind and sea spray. The fish pieces lay some twenty paces away amid seaweed close to tall slabs of rock at the high-tide mark, and I had added a garnish of peanut butter – irresistible rodent bait. Now all I had to do was wait.

An hour passed and the light was beginning to dim, when something moving from left to right at the top of the rocks caught me by surprise: a tantalising glimpse of tail disappearing between clumps of sea pink. I held my breath. Could this be one? There are no land mammals other than black rats on the Shiants, so it had to be. Please let me see you, little fellow, I whispered to myself. I was downwind and motionless, though close enough to have been spotted. Would it dare break cover for the sake of a free meal? Nothing stirred for a while, until . . . There! It sprinted to a new hiding place well above the bait – a rat, no mistake. Several minutes went by before it appeared once more, scampering down the face of the rock and dodging behind a piece of old tarpaulin lying next to a rusty boat winch. Black rats are known for their climbing skills, nevertheless the ability to descend slippery vertical stone head-first at such speed, without falling, was extraordinary.

The rat was now a yard from the strong-smelling bait, and I knew it couldn't stay concealed for long. Sure enough it darted out, grabbed a mouthful and slipped back into a rock crevice. During those few seconds I was able to get my first decent look, and could see the differences that distinguish it from a brown rat. It was less heavily built and more pointed, with larger ears and shaggy, charcoal-coloured fur. And at that moment a second arrived, scurrying along the base of the rock to the food. Their colours can vary, and this one was much greyer. It took a morsel of peanut butter and hid, and they peered out towards me, trying to figure out the imposter on their beach. Bit by bit they gained in confidence, until I was able to watch them sitting in the shadows beside the food, tucking in greedily. Strange to say, they were actually quite cute.

There is no record that I am aware of for the 'longest journey in Britain to see a rat', but I feel fairly confident that my trip from Dartmoor to the Hebrides could lay claim to the title. (And I don't imagine I face much in the way of competition.) This was a thrilling sighting like no other, in a place like no other, and when it became too dark to see any more, I returned to the bothy, grinning from ear to ear, and enjoyed a triumphant dram of whisky.

Overjoyed as I was to see such a British rarity, I did, admittedly, have trouble sleeping that night. Alone in the bothy, lying in my sleeping bag, I kept thinking I heard scratching and scuttling footsteps above me, below me, at the other end of the room. I was tormented by ratty nightmares.

The next day the weather was stormy, as forecast, and I woke early to the jazz drum-brush rhythms of rain on the window panes. I lit the fire and got the kettles boiling, and everyone drifted in for breakfast before setting out for another day's ringing. The rain cleared quickly and I spent time with the BBC crew talking about the rats, and joined them on a hazardous scramble around the coast of Garbh Eilean to film puffin chicks – fluffy balls with the wonderful name 'pufflings' – narrowly avoiding being cut off by the tide on our return. I was also keen to explore the main island and accompanied trip organiser Jim on a walk to the southernmost point to count birds. Childhood holidays in 'wild places' had shaped his lifelong interest in the outdoors, he told me, and the Shiants had made a profound impression ever since he first visited.

'I love islands, and birds, and ringing, and it offers all three,' he said. 'The remoteness and simplicity has its own challenges, though. People might consider coming somewhere like this to be escapism, but if anything it's tougher because you can't run away from yourself here. It can leave you refreshed, or your emotions raw.' He smiled. 'That's when the malt in the cupboard helps. Either way, you're usually pleasantly knackered by the end of the trip.'

We were watched on our walk by two great skuas, which hung on the wind above the high ground. No bird I know matches their mesmerising aura of self-confidence. Large and brown, these broad-chested pirates, which rob other seabirds of their fish catches, are quite simply as tough and tyrannical as they come. One had a metal loop fixed on the upper part of its right leg, indicating that it may have been ringed in Scandinavia, by someone brave enough.

Without a watch I had no sense of the time, and no real need to know it. Morning drifted into afternoon into evening, and after supper I spent dusk down on the causeway once again. The rats were back, and bolder than before, so I experimented by moving much closer to the peanut butter bait in order to get a photo. One ran out, and as it grabbed a mouthful I pressed the shutter button, the flash went off and it sprinted back under a rock. You would think that would have been an end to it, but back it came half a minute later, turning tail whenever the camera fired. A second joined it, then a third clambered down a boulder opposite me and across one of the old winch ropes, and a fourth ran from my right a few feet away, a little too close for comfort. It was time to take a few steps back and leave them to it.

Over the two days that followed I enjoyed spending time exploring on my own and in the excellent company of the Shiants Auk Ringing Group. You had to admire their courage in scaling the slopes to get to the bird colonies, and I watched from a distance as they used traditional fleyg nets with long handles to catch flying razorbills and guillemots. I also shared the novel experience of tape-luring for storm petrels at night, which involved broadcasting their peculiar purring call in the hope of drawing the birds close, and I was fortunate enough to see one flickering past me in the half light, evading the waiting mist net and disappearing. Getting into the spirit of things, I even cooked everyone an evening meal, which I don't think had been factored into the group's risk assessment, and took the opportunity to toast the rats and seabirds. Not literally, of course.

Marooned on an island, you get to know people quickly, and I was sorry when the trip drew to a close and we waved goodbye to the BBC crew and prepared for our own departure. Everyone worked hard to leave the place as they had found it, and the human chain that formed in order to pass bags down to our boat the next morning was as symbolic as it was efficient. I felt lucky to be a part of it.

Of course the group couldn't resist logging a few more birds, and on the journey back to Skye we stopped off at a little uninhabited island called Fladaigh Chuain, where green-eyed, ragged-black shags, beaks gaping, necks snaking, honked and hissed at us from their reeking nests as we clambered ashore. The aim was to ring arctic tern chicks, and once we had reached the nesting area on open ground in the centre of the island everyone was instructed to spread out in a line and tread carefully – the tiny young were well camouflaged and it would have been easy to tread on one. We donned caps to protect our heads from the dive-bombing parent birds and moved fast in a single sweep through the screeching colony. I was offered the chance to ring a chick, and was talked through the process, squeezing the metal clasp shut around its leg with pliers before gently placing it back on the ground. It is the first and only bird I have ever ringed, and what a fabulous species to start with. Destined to undertake the longest migrations of any animal on the planet, my tern, number ST12266, would be heading to Antarctica and back, year after year, all being well. I wished it safe passage on its travels.

Five days on the Shiant Isles had provided a lifetime of memories, as well as unexpectedly good views of black rats, but it was time to head home – back to reality, or perhaps back from reality. Unlike those of the young arctic tern, my travels were nearing an end, with just a couple more sensational scarcities still to see.

4.45am. I slapped my alarm clock silent and lay staring up at the ceiling, debating whether to get up. It had been a long week at work, and I'd treated myself to a couple of cans of beer the night before. Of course, I only bought a six-pack so

I could ensure the plastic rings were cut through to prevent turtle deaths, and drinking the contents seemed the most sensible way to dispose of the alcohol safely. But doing my bit for the environment didn't help when it came to an early start, and I rolled over and began to doze off. Why bother trying again, I sighed. I had already been to Studland Bay four times to look for seahorses without luck, and even though two pairs had now been located, it just seemed like, for me at least, it wasn't meant to be. Giving up was the obvious course of action.

Only, I really wanted to see one, and that wasn't going to happen lying in bed. I couldn't bear the thought of getting an email later on saying the dive had been successful. If in doubt, go – that had become my motto. I hauled myself upright. One final visit, I decided, and I got ready and set off, driving east towards the rising sun.

On my last attempt to find a seahorse, I snorkelled alone. However, this time I would join Neil Garrick-Maidment and Seahorse Trust volunteers once again. It was forecast to be a beautiful August day, the prevailing wind direction was good, and the omens also seemed promising: approaching Studland, I passed a plumber's sign with a seahorse design on it, and on arriving I parked near a brewery van with a similar illustration.

Neil pulled up in the car park along with two other divers – John, who I had met on a previous survey trip, and Eva – and once they got kitted out we walked down to South Beach and waded out into the sparkling sea. The beach was empty, the setting was glorious, and the water was clear and a pleasant 19°C. No place better to be – not even a warm bed. After final checks, the three of them submerged and I followed Neil's surface buoy out towards the seagrass, spotting a crab and a few colourful wrasse on my way. Marine life was out and about in the sun.

I swam ahead of the group and explored the shallower areas, resting every now and then to rinse out my face mask and assess the direction of their buoy. Every time it paused, my heart beat faster – perhaps they had found something. Then it would shift forward again, rocking over the waves, and I carried on.

After three-quarters of an hour the trio surfaced. 'Nothing yet,' Neil said, checking his position relative to the shore. 'Something large swam by though, perhaps a seal.' And down they went again, continuing back along the edge of the seagrass. Carrying only enough air in their tanks for an hour-and-a-half's dive, they needed to strike lucky soon.

I saw a pipefish, and a couple of bass hiding among the green fronds, but hope ebbed away with the passing minutes. Despite all my efforts it was obvious I was never going to come across a seahorse, and I decided it was probably a mythical creature in any case, a unicorn of reefs and bays dreamed up by naturalists and divers with vivid imaginations.

At that moment, Neil came to the surface a little way off and called over in a calm voice: 'We've got one.'

I could hardly believe it.

'A male spiny,' he said, beckoning me over.

I started swimming as fast as I could, except that as soon as I kicked out, my calf muscles cramped up, just like before. Not now! It was ridiculous. All I had to do was cover twenty yards and there I was, floundering like a novice swimmer, legs buckled in pain, desperately pulling myself forward with my arms.

Teeth clenched, I finally made it to Neil and managed to straighten my calves until the cramp eased, then looked down into the water beneath him. A little sediment had reduced visibility, but I could see Eva's fluorescent yellow fins and the murky forms of her and John at the bottom – quite a long way below. I pulled my snorkel to one side, took a deep breath and swam hard down through their rising bubbles until I was right above them. They were facing towards the seagrass, only I couldn't pinpoint what they were looking at and had to hurry back up to the surface, gasping and spitting out seawater. I realised this wasn't going to be easy. A lungful of air and a buoyant wetsuit made it hard to submerge; I was rushing, and I needed to relax in order to hold my breath for longer. I tried again, and got a bit closer, but still couldn't see . . . ah, there was something. Was that it? A thin shape, quite dark among the green fronds, crowned by soft spines . . . Yes!

I had to ascend again, grabbed another breath of air and ducked under, pumping my flippers to reach the three of them. Eva held out a hand to help me stay down, and although the seahorse was mostly hidden by seagrass, I got a better view of the side of its head, before I was forced to race back up again. It was incredible they had spotted it at all.

My lungs were aching. John came up and kindly offered me his weighted diving belt, however I felt a bit nervous, having never worn one before, and decided to persevere without. Again I took a deep breath and dived, following Neil's buoy rope down through the cloudy surface to the clearer water below. This time I approached the group from the other side and Neil extended his forearm so I had something to cling to. Now I had an excellent view. Wow! The well-camouflaged seahorse, about a handspan in length and looking remarkably calm, was upright and side on, its pointed pipette snout visible and its tail running across a short tangle of seagrass to the right, where the tip curled around a bent stem. This was no mythical creature – although our wildlife doesn't come much rarer or more mysterious. This was real and unforgettable, and such a close encounter repaid all the effort. I could have reached out and touched it.

Neil took a couple of photos for identification purposes and, after a final look, we surfaced and returned to the beach, which was filling up with people. Walking back up to the cars, we passed through a wedding party and stopped to watch the Red Arrows performing a display over Bournemouth – it seemed like the whole world was celebrating.

I was sorry to say goodbye to Neil and his dedicated team, and couldn't thank him enough before starting off on the long drive home, shaking my head with a smile in disbelief – I had done it, he had done it, they had done it: actually found a wild British spiny seahorse. Amazing. And to think, I almost hadn't come at all.

The next day Neil emailed me two of the photos he had taken of the individual, which had been given the reference number E457. 'Perhaps we should call him Charlie,' he wrote. That would be an honour indeed.

I had missed seeing the llamas, a herd of them winding up steep trails to a tarn set high among the Cumbrian hills. Llamas are not exactly a common sight roaming the uplands of Britain, but that doesn't mean that in a moment of madness I'd included them on my list. No, it wasn't the Andean pack animals I was interested in, it was the contents of their saddlebags, and I was over two years too late. I only wish I'd been there to witness one of the more unusual and ingenious rescue efforts ever mounted to save a species on the brink of extinction in Britain.

The Lake District, playground for ramblers, poets, landscape-lovers and, it seems, the occasional llama, owes its natural beauty to the actions of ice on rock. During the last deep freeze, massive glaciers, born in bowl-shaped corries where snow accumulated on the mountainsides, ploughed down steep-sided river gorges, transforming V-shapes into U-shapes, creating narrow ridges between the widened valleys and gouging myriad hollows, which filled with water as the ice melted. Like the clear lochs of Scotland, these lakes and tarns, some plugged by dumped glacial debris, became home to isolated populations of fish, including two secretive species: the schelly (also known as the powan or gwyniad) and the vendace. Plain and silvery whitefish, they look much like herrings, but are actually related to salmon and bear the family's signature adipose fin – a small, nub-like appendage on the lower back between the dorsal fin and tail. Of the pair, the scarcer vendace is considered the rarest of our freshwater fishes.

Only four native populations of vendace have ever been recorded in the UK: two in south-west Scotland and two in north-west England, and those in Scotland became extinct decades ago. Now confined to the Lake District, this relic of the ice age needs cool, clean, well-oxygenated water in order to survive. Pollution, rising temperatures, predatory fish and the effects of weed and silt on gravel beds where females lay their eggs have spelled disaster. By 2001, repeated surveys at Bassenthwaite Lake began to draw a blank, leaving just a single precarious population in Derwentwater, and this would be its sole sanctuary but for the efforts of quick-thinking conservationists. Offspring were taken from Bassenthwaite in the nick of time and a refuge population established at Loch Skeen in Scotland. In addition, young raised from Derwentwater eggs were placed in watertight containers and carried to remote Sprinkling Tarn several miles upriver to be released. Such an inaccessible site presented a transportation challenge for those involved at the Environment Agency, and the answer was provided by a local llama-trekking organisation, whose sure-footed herd conveyed the precious cargo up stony paths to their destination in 2011 – a vaguely comical spectacle with a serious objective.

Having missed the translocations, my options for seeing a wild vendace were extremely limited. They live in the darkest depths of lakes, so snorkelling was out of the question, even taking into account my six-foot plunge-diving abilities.

Instead, all hope hung on a licensed survey being carried out by public-sector research body the Centre for Ecology and Hydrology. Two experts were due to spend a single night in September monitoring fish diversity in Derwentwater using six static nets and echo-sounding equipment, and I arranged to meet up with them.

After a slow motorway drive north to Cumbria, stuck behind cars towing caravans and motorhomes towing cars, I reached the Lakeland town of Keswick and located my guest house accommodation in a neighbourhood where every geranium-festooned property was a B&B. Front doors and windows were plastered with B&B signs, and B&B boards stood in rows in front gardens. You could earn a good living in the area by simply setting up a hardware shop specialising in selling the letter B.

I wandered around the market town centre and its hiking-gear shops, though didn't have time for the local Pencil Museum – I'll pen it in my diary for another visit – then made my way to the edge of Derwentwater, walking down past the lakeside theatre to wooden rowing boats lined up by the jetties before following a path through woodland to a rocky headland known as Friar's Crag. Here I stood and gazed out at the magnificent view across the water to rugged fells beyond.

An information board beside the track described the islands and local wildlife, and I was interested to see it highlighted the fact that the lake is home to our rarest freshwater fish. Very few people have ever, or will ever, see a vendace in Britain – including anglers, who are prohibited from catching them in any case. Yet even though vendace might as well be invisible, visitors must look out from the sign and feel, as I did, a strange thrill at knowing the water hid something special. Rarity captivates us. The notion has a potency that extends beyond a simple expression of quantity. It resonates with significance, a sense of shared infatuation, and touches on a range of emotions. I certainly felt its pull. Somewhere far below the surface of the lake, I thought, vendace would be feeding in the deep channels they have called home for thousands of years. I also realised this might well be the closest I ever got to them.

By the time I returned to the jetties, fish ecologists Janice Fletcher and Ben James had arrived and were unloading their apparatus into a boat, and I helped carry equipment down before waving them off. They had a long night ahead gathering echo-sound data of fish shoals, and I would join them in the morning to lift the nets.

It wasn't easy getting to sleep that evening for wondering whether the vendace were stirring and we would have any success. I woke early and needed a full cooked breakfast, leaving the street's heady smells of frying bacon and floral displays behind and driving once more to the lakeside. There wasn't much room on the boat, but Janice and Ben were good enough to make space for me. Both of them had been involved in surveying Derwentwater over a number of years, and said that a few vendace typically turned up in the deepest of the six nets, but numbers had declined. They also warned me that prolonged warm weather over the summer might have altered oxygen levels in the lake and forced them to move into higher water away from their usual haunts. Don't get your hopes up too much was the message.

Ben steered us out past Derwent Island to the first pair of buoys, and Janice donned gloves and began hauling in the nylon netting. I was astounded at the number and size of fish caught, in particular large roach and hefty perch. No vendace though, and we motored south to the next buoy. Apart from making a note of the net number and time, there wasn't much I could do to be useful, so I just chatted away, trying to appear composed when I was actually churning inside with tense anticipation.

The second net brought up more perch and roach, a pike and a small ruffe – a species that has recently found its way into the lake and poses an added threat to vendace by competing for food and eating its eggs. Once again the rarity was not to be seen. Nor in nets three or four.

'The next net is the deep one,' Ben said, as we approached the centre of the lake.

At twenty-two metres this wall of mesh, the height of a six-storey tower block, extended to the lowest point in the lake and had been set specifically for vendace. It took an age to lift and I was tempted to help Janice, but thought better of it. One person at a time standing up in the little boat was enough, so I sat watching flickering blades of light rising in the water as perch and roach came to the top, willing the rarity to appear.

'Nearly there,' she said.

Metre by metre, hand over hand, the net emerged, until the last section to be heaved from the depths broke the surface, I held my breath and . . . nothing. That was it. The lake was keeping the secretive species of its glacial past hidden deep within in its belly, and the best chance I would ever get to see a vendace had ended in failure.

'Ah well, not to worry,' I said, trying not to look too downhearted. My good luck had been destined to run out, but it didn't make it any easier. Every long trip with an uncertain outcome is demanding both physically and emotionally, and, much as I put a brave face on it, I felt utterly deflated.

There was only a six-metre floating net left, which had been positioned fairly nearby over the deep water, and Janice pulled it in as we prepared to head back for Keswick. If the vendace had moved significantly higher off the bottom there was still a slim possibility of getting one, and scarce species had certainly toyed with me before, stretching the suspense until the bitter end. However, this would be asking for too much: the final net of my final trip of the year.

A few wriggling perch and roach came aboard as the netting was lifted, and I sat back, taking in the scenery and considering the drive back home.

'Here we go,' Janice said calmly.

'What?' I quickly lent forward and watched as a streamlined fish with a faintly olive-brown back was hoisted up the side of the boat within the tangle of nylon and laid on the pile of mesh. It couldn't be! And yet, there it was, a vendace, shining in the sun, beautiful and ...

'There's another coming,' Ben added, and a second came aboard within the tail section of netting. They had been right to suspect warm weather might lift the fish off the bottom, and fortunately the lowest part of this floating net had been just about deep enough to catch a couple. Not only was it the final net of my final trip of the year, but it had actually been the final metre of the final net of my final trip of the year.

Both vendace were of a similar size, around eight inches long and perhaps two or more years old, and in prime condition. Their delicate scales were neat and smooth, their pupils large and dark, and there was something very satisfying about their straightforward shape and unfussy appearance. They were a no-frills fish of simple elegance. And it was safe to say that these were the only vendace anyone had seen in the country since previous sampling a year earlier. It was a genuine privilege to set eyes on them, as if long-lost bars of silver, cast into the depths millennia ago, had been raised from the bed of the lake, and to hold one was to touch the past.

On the drive home I had to restrain myself from telling complete strangers in motorway service stations about what I had seen, and managed to do the same with colleagues at work the next morning. Hard as it was, I kept my excitement at encountering such a natural treasure to myself (surprisingly, not everyone wants to know about rare fish you've come across). I bottled up the memory, but was treated to an unexpected reminder a few days later when walking the dog. Something caught my eye, and looking down at the sleeve of my coat I noticed a couple of scales still stuck to the material – vendace scales, glittering like tiny fragments of ice.

# Twelve

I was used to searching, but this time I didn't have a clue where to look. I had been walking for a while, and now retraced my steps along the winding route I had already taken, pausing every so often to raise and lower my gaze, hoping that what I sought would catch my eye and make itself known. Not for the first time, I was stumped when it came to identification by a lack of knowledge, and eventually I stopped, checked the piece of paper in my hand and realised I needed help.

'Excuse me, I'm sorry to trouble you,' I said to a passing member of staff, 'but I can't remember, is pancetta a cheese or a kind of bread?'

'Aisle three,' she said. 'Cold meats.'

'Thanks,' I replied, and pushed my trolley on through the supermarket, feeling like I had made the kind of basic blunder that would have me hounded out of the middle classes by outraged foodies brandishing Carluccio cookbooks and hurling handfuls of fusilli. Still, I thought, at least while I'm in the meat aisle I can pick up some mascarpone and brioche before heading to the wine department for prosciutto.

I had decided to cook a special family dinner to celebrate a successful year and the near-completion of my rare species quest, but was beginning to regret picking a recipe on the basis of a glossy photo in a fancy European cookery book. The endless costly ingredients added up to more than a meal out, and although I wanted to thank my wife and daughters for their patience over the preceding months, they needed plenty more, as it took until 10pm before the food was ready. Burnt bits aside, it almost resembled the picture in the book if you stood several yards away from the plates and squinted, which I regarded as a personal triumph.

Autumn had crept in on lengthening shadows. Worm casts peppered the lawn, jays appeared as if from nowhere and damp air misted windows. As temperatures fell, low clouds refused to budge from Dartmoor, and some days it was so foggy on morning walks that I could hardly see the dog at the end of his lead. With no trips planned for a while, this was a time to reflect on all the rare animals, and rare moments, I had experienced and to savour memories of exciting sightings and unforgettable places.

I'd had no idea, three years earlier, just how far a trail of paw prints in the snow might take me. Following tracks to catch tantalising glimpses of Scottish wildcat in the wintry Cairngorms had marked the start of a journey that would eventually cover the length and breadth of the UK. While wildcats declined a second audience, the other mammals I had been lucky enough to see included delightful dormice on the Isle of Wight, a brief close-up view of a pine marten in the forests of Scotland, an elusive Bechstein's bat in a Dorset woodland and black rats on the Hebridean Shiant Isles. The five fish had begun with monster common skate off Oban, followed by European eels caught in Norfolk, majestic basking sharks feeding in the waters of the Isle of Man, a surreal spiny seahorse at the fifth time of asking in Dorset and vendace in the Lake District. Reptiles and amphibians ranged from great crested newts found on a freezing night near Peterborough and a heathland smooth snake on the south coast that bit me, to fast-moving sand lizards in Dorset, croaking natterjack toads in Cumbria and pool frogs at a secret site in East Anglia. And the invertebrates I sought started with the handsome Duke of Burgundy butterfly in Sussex, then the streaked bombardier beetle that almost evaded detection in London's Docklands a month later, before good views of Norfolk hawker dragonflies and the impressive fen raft spider, and finally a queen shrill carder bee in Wales. A little of what the United States military refers to as 'mission creep' meant that I also added a hunt for the outlandish wart-biter cricket to my travels, as well as two nights on a mountain looking in vain for the rare Silurian moth. As for birds, I was blessed with a song and fleeting sighting of a golden oriole in Suffolk and the chance to stroke a corncrake on the island of Coll, and saw breeding Slavonian grebes in the Scottish Highlands and a pair of Montagu's harriers in Norfolk. I had a single bird still left to see to complete the challenge I had set myself.

One of the unexpected highlights had been all the other scarce and notable animals spotted along the way, including a sperm whale, a displaying capercaillie, bitterns, black-throated divers, a lesser-spotted woodpecker, stone curlews, an Alcathoe bat, brown-banded carder bees, silver-spotted skipper butterflies . . . the list goes on. And of course it was wonderful to chance across red squirrels, water voles, otters, adders and all manner of other fascinating finds, many of which I would never have seen otherwise.

I had also enjoyed meeting so many friendly and enthusiastic naturalists, who guided me up steep learning curves and took the time to show me the precious variety of life on our doorstep. They had taught me that the more you look, the more you see; the more you listen, the more you hear; and the more you learn, the more you understand. I had plenty of looking, listening and learning ahead, and I also wanted to do a little more to help. Every species in decline or living on the brink in our country tells its own story about the environment and, ultimately, ourselves. Natural history is our history, from the early felling of trees, drainage of wetlands, reclamation of heaths and enclosure of fields to the protection of game, planting of conifers, intensification of farming, expansion of cities and creation of reserves.

While some native animals have adapted to change, others, particularly those with more exacting requirements, have suffered. We have a lot worth cherishing, and unfortunately the importance of conserving biodiversity is seldom high on the political agenda. Our present preoccupation with economic growth, coupled with the demands of an expanding population, mean that national, as well as global, targets to halt species extinctions keep getting kicked into the long grass – and the loss of meadows means there isn't much of that left. The fact that there are conservation charities striving to save our bats, birds, bugs, bees, butterflies, basking sharks and everything else besides is extremely reassuring, and they deserve our support. An uncertain future faces many rarities, and even widespread species can't be taken for granted. The great auk egg that I held at Oxford University and the tiny Ivell's sea anemone examined under a microscope had been reminders that there is no bouncing back when numbers hit zero.

After the cold spring, a warm and dry summer had been good news not only for me, but also for some of the species I had encountered, such as the Duke of Burgundy butterfly, though many others remained in a perilous position, including spiny seahorses, which totalled just four in Studland Bay. On a positive note, a young vendace was unexpectedly caught by Dr Ian Winfield, of the Centre for Hydrology and Ecology, during an autumn fish survey at

Cumbria's Bassenthwaite Lake – the first to be found there in more than a decade. And the male basking shark I had watched being tagged by the Manx Basking Shark Watch crew, named Fricassonce, was tracked by satellite over a period of five months, swimming north before the tag popped off and was recovered in November, its data helping map the movements of these mysterious giants.

My adventures were nearly at an end, yet I would happily have started over once again, seeking new scarce and threatened species or revisiting those already seen. What keeps wildlife encounters so stimulating is that you never have the same experience twice. However, I was looking forward to slowing down, spending time at home and getting to know more of the wildlife where I live. Searching for rare animals had, paradoxically, taught me to better appreciate the common. I was ready to work my way back from the few to the many, from the once-in-a-lifetimes to the everyday, whether dragonflies, moths, newts, bumblebees or butterflies. Sure, I would struggle to name half of what I saw, but noticing was a start.

Winter kept the weather forecasters busy as storms hammered the country and caused widespread flooding, and it took until spring before the conditions settled down. I bided my time and waited for my final scarce bird to arrive. I didn't know when or where it might turn up – nor did I know exactly what it would be. That might sound a bit odd, but I had a selection of unusual migrants in mind: I was after a rare breeding visitor that pointed to the future rather than the past.

Being so mobile, it's hardly surprising that birds get around a bit, and foreign oddities are constantly touching down here, especially given our position at the western edge of Europe. Mostly these travellers either move on or die, and are largely insignificant in ecological terms. However, a few manage to get a scaly toehold, gradually becoming a part of our recognised fauna, and while the total population of birds in

Britain has fallen, the variety of species is increasing, with warmer weather enabling more southerly breeders to spread north. I was hoping to see an avian curiosity that might eventually colonise our shores, and May 2014 brought just the rarity that I sought.

Black-winged stilts are one of the world's most elegant and eye-catching birds, with glossy dark upperparts, delicate straight bills and preposterously long, reddish pink legs. They are widespread waders of the kind one spots paddling around waterholes on TV documentaries about African wildlife or picking at insects in the shallows of southern European marshlands. Yet they do venture further up the globe, adding to the multinational mix of visitors found stalking the edges of our lakes and estuaries, such as spoonbills, glossy ibises and cattle egrets. And occasionally they stay awhile. One individual famously settled at Titchwell Marsh reserve in north Norfolk in 1993 and remained there for more than a decade. The lonely male, who never met another of his kind in all that time – and in desperation even tried to mate with local oystercatchers – became probably our most watched wild bird ever before he disappeared in 2005. He had a nickname, too, which adhered to the alliteration custom in such situations: Sammy the stilt.

Now and again, black-winged stilts nest in Britain, and the signs point to the possibility that they might one day become established here. So, in May, when several abandoned the Mediterranean during a prolonged dry spell and arrived in south-east England, the hope was that they might pair up – and not with oystercatchers. To the delight of onlookers, clutches of eggs were laid at Cliffe Pools in Kent and at the newly established Medmerry reserve near Chichester in West Sussex, which is where I headed to try to see one. I was confident I should get lucky – all I had to do was find the reserve, then the nest, and there should be a stilt sitting on top of it.

It wasn't quite as easy as I imagined. Medmerry reserve was still in the process of being created as part of a coastal realignment scheme, and I had been advised to park at the RSPB's Pagham Harbour visitor centre and walk. Only, by the time I got there in the late afternoon, the centre had shut and there were no staff around who could give me directions.

'You haven't seen a stilt have you?' I asked a couple returning to their car.

'A what?' the man said, looking puzzled and turning to his partner.

'What's he lost?' the woman asked him.

'Never mind, thanks anyway,' I said, and tried a few other visitors without success, before getting back into my car and driving to Earnley, a little closer to the area where I understood the birds were nesting. I parked by the village church and walked along a road through farmland and beside holiday chalets until I reached shingle banks at Bracklesham Bay. A lack of birdwatchers had me worrying whether I had got my bearings seriously wrong, and when a dog walker came up the track I took a chance he might be able to help.

'Sorry, this might sound like a strange question, but have you seen any birdwatchers around here?' I said.

'Yes,' he replied with a grin, and walked on.

'But . . .'

He stopped, and turned with a laugh. 'Oh, you want to know where they are?' A joker. Then again, I was willing to take mild teasing in exchange for information, and he pointed the way down a path to a metal container at the far end. 'And don't worry, the stilts aren't going anywhere,' he called behind him as he headed off.

The metal container turned out to be a portable cabin for RSPB volunteers providing the star attraction with a twenty-four-hour guard, and the nest site had also been surrounded by an electric fence. No way were foxes or egg collectors going to ruin this reserve's good-news story. I introduced myself to the two volunteers standing by the shelter and they pointed to a narrow muddy spit at the centre of a flooded area opposite, and I set up my telescope and tried to locate the spot in the viewfinder: sky . . . down a bit . . . more sky . . . water . . . up a bit . . . mud . . . stilt! There it was, a female black-winged stilt sitting on her clutch. The nest was little more than a scrape in the earth between stones at the water's edge, and I was told she had three eggs that were about a week away from hatching.

She was lovely looking, but it's the legs that make a stilt special, and they were tucked like bent straws beneath her. However, ten minutes later the large, darker-backed male arrived to swap incubating duties and I got an excellent view of the pair standing side by side. They looked to have been assembled from kits that included wrong-sized limb pieces. While their heads and bodies were smart and proportionate, their spindly legs were impossibly long. It appeared as if they were floating above the ground, tethered to the mud by lengths of red string. Yet they moved with such grace and refinement that it all made perfect sense.

I watched the female feeding, wading deep and picking insects from the surface, and later the male came off the nest and flew overhead with legs trailing before landing on a muddy expanse amid avocets and little-ringed plovers. It was a peaceful scene on a summer's evening, with few people about, and I sat and listened to the sounds of birds and gentle sighing of waves drifting in on the warm breeze. This nearly completed Medmerry sanctuary of brackish lagoons and muddy creeks was the result of a major engineering project, which involved moving tidal defences inland and surrendering overflow areas to the sea to reduce the risk of local flooding. Built for the century ahead, it was appropriate that the new reserve should be christened by a rare bird that could well become a part of our future. This pair of black-winged stilts, looking perfectly at home in West Sussex, made for an uplifting conclusion to my travels – and I later heard that all the eggs hatched safely, both here and in Kent.

Time, then, to pack away my maps, clear the accumulated rubbish from my car, put the midge repellent and seasickness pills back in the bathroom cupboard and call it a day. Except that something was troubling me, fluttering around my thoughts, and I couldn't bat it away. I wanted to see a rare Silurian moth, and I hadn't. While every other animal had been good enough to show itself at some point, this small, brown insect had given me the slip. I might easily have dismissed it with a shrug of the shoulders. It was no showy or spectacular species after all, just a moth, and I had already exceeded the year I originally scheduled for trips. Yet there was something about it that embodied all that was exciting about rarity. Flying late at night, high in the Black Mountains of Wales, this was a creature that refused to make life easy, and only a handful of committed naturalists had ever seen one. The spirit of wilderness borne aloft on little wings. It was a challenge I couldn't resist, and I had one more opportunity to try to track one down. Expert entomologist Dave Grundy, whom I had accompanied before, was planning a return visit to the Hatterrall Ridge on a single night in July, and it was likely to be the only survey for the foreseeable future. I contacted him and said to count me in.

The date finally arrived, with the weather set to be far kinder than the previous year, and four of us gathered at the base of the ridge before setting off on the steep climb to the summit. With myself and Dave were Chris Manley, a wildlife sculptor and moth book author, and Chris Williams, an energetic moth enthusiast aged in his seventies, who had been on the 2013 expedition. It was warm, clear and calm, but no one dared get their hopes up too much, and once at the top we worked hard until nightfall, setting up various light traps in promising areas of bilberry. Moths were certainly on the wing, though whether the fickle Silurian was among them was another matter. There was nothing to do but wait. And what a magical place to kill time, far from city lights, two thousand feet up under a sky full of stars.

Head torches were eventually switched on and we began checking the traps, which had attracted a good variety, even though it was only 11pm. Moths included true lover's knot, smoky wave, scarce silver Y and small elephant hawkmoth. And although it is generally wise to keep your mouth shut in the company of experts, I did risk blurting out a few names and accurately identified a brimstone, mottled beauty and flame shoulder moth – plus points that were subsequently cancelled out by misidentifying three others.

I was kneeling in the heather inspecting finds with Dave when Chris Williams walked over from one of his light traps holding a small transparent container.

'I've something to show you,' he said, handing it over to Dave. I could see he was restraining a smile.

Dave took one look and beamed. 'There you go,' he said, and passed me the plastic pot. 'A Silurian moth.'

Already? Surely that wasn't possible! Not only are Silurian moths supposed to fly after midnight, but they were also meant to keep me hanging on until the last trap check of the night. Hadn't they read the script?

I examined the moth using my torch. It was short and compact with a furry body, and its reddish brown wings, held together over its back, were modestly marked with a faint wavy band. This neat and distinguished species was the exclusive prize that heralded the completion of my journey, and I clutched the container as if it were a medal awarded at the finish line of a marathon. It was fantastic to see one finally and I was glad I had made the extra effort to be here, sharing in the simple enjoyment of our natural diversity with others. After all the travelling, I felt I had in some way arrived, and it was a quietly moving moment.

This was to be far from the only Silurian moth of the night. Another was found in a light trap tucked in a sheltered spot on the slope, and then two in a trap set up in dense bilberry, and two more further along the ridge, and they kept on coming. At one point, I sat alone beside one of the lights as three or four flew around me, and I was able to coax calmer specimens onto my outstretched fingers. Dave had never seen anything like it before. The timing and conditions had proven perfect, and as the hours passed, the number of Silurians in some traps began to reach double figures, outnumbering typically common species. This was a rare night – a night of the rare. By dawn, a record-breaking 126 Silurian moths had been logged, and in areas of the ridge where they had never been found before. It was a significant chapter in the story of this enigmatic species, and I felt very fortunate to have been a part of it.

Before we packed up and prepared to walk back down to our cars, I stood alone to greet the new day, looking out from the edge of the ridge over Herefordshire and towards a wide, cloudless sky flushed pink by the rising sun. Mist hung in the valleys and over villages where early risers would just be getting up. Across the land spread out before me and beyond the horizon to the seas that surrounded Britain, animals would also be stirring – the day shift clocking on as night dwellers retreated into the shadows – and I was able to imagine precious rarities among them: a smooth snake seeking out a warm patch of light; a seahorse displaying amid a tangle of seagrass; a Norfolk hawker dragonfly taking to the air on rustling wings; a golden oriole flooding a woodland avenue with song; a Scottish wildcat retreating to its lair; a basking shark breaking the surface in a tranquil bay. And on this day, someone, somewhere, might just be lucky enough to chance across one of them.

# Acknowledgements

I am immensely grateful to all those who helped make this book possible. One of the pleasures of wildlife watching is sharing the experience with other people, and I enjoyed enormously the company of experts and enthusiasts who gave freely of their time and enabled me to track down so many incredible species.

Special thanks to commissioning editor Julie Bailey at Bloomsbury Publishing for her enthusiasm, Susan Smith at MBA Literary Agents for all her support over the years, and copy editor Rachael Oakden for her excellent work. Also to my wife, Nicky, and daughters Emma and Beth, to my parents, Andrew and Penny, and Pat, as well as my brother, Jay, and sisters Charlotte and Tamsin. In addition, a special mention to Derek Niemann and colleagues at the RSPB, and to the Society of Authors, whose Authors' Foundation award helped fund my trip to see a Scottish wildcat in 2010.

My thanks go to the following for their help, information and encouragement, with details taken at the time of writing; I apologise if I have unintentionally left anyone out.

Ivell's sea anemone: Professor Richard Ivell, Leibniz Institute for Farm Animal Biology; Dr Sammy de Grave, Oxford University Museum of Natural History.

Common Skate: Dr Viki Wearmouth, Marine Biological Association; Plymouth City Museum and Art Gallery; Ronnie Campbell, *Laura Dawn*, Oban; Francis Neat, fish biologist, Marine Scotland Science; anglers David Griffiths of fishingbooksender, David Morris, John and Paul; Jane Dodd, marine operations officer, Scottish Natural Heritage; Steve Bastiman, Scottish Sea Angling Conservation Network.

Great crested newt: Paul Furnborough, conservation officer, Froglife; Rebecca Neal, conservation youth worker, Froglife; Kathy Wormald, chief executive officer, Froglife.

Smooth snake: Rob Farrington, RSPB visitor manager, Arne; Michael Wilson, Arne RSPB reserve.

Bechstein's bat: Colin Morris, nature reserves manager, Vincent Wildlife Trust; Daniel Whitby, consultant ecologist, AEWC Ltd; Natalie Buttriss, chief executive officer, Vincent Wildlife Trust; Helen Miller, Bat Conservation Trust; Lizzie Platt, Devon bat warden.

Sand lizard: Tony Gent, chief executive officer, Amphibian and Reptile Conservation.

Duke of Burgundy butterfly: Neil Hulme, chairman and conservation officer, Butterfly Conservation Sussex; Dan Hoare, South East senior conservation officer, Butterfly Conservation.

Hazel dormouse: Ian White, dormouse officer, People's Trust for Endangered Species; Jill Nelson, chief executive officer, People's Trust for Endangered Species.

Golden oriole: David Rogers, RSPB site manager, Lakenheath Fen.

European eel: Peter Carter, Norfolk.

Natterjack toad: Neil Forbes, National Trust ranger, Sandscale Haws National Nature Reserve; Tony Gent, chief executive officer, Amphibian and Reptile Conservation; Norman Holton, senior sites manager, RSPB Cumbria coastal reserves.

Corncrake: Ben Jones, RSPB reserve warden, Isle of Coll; Professor Rhys Green, RSPB principal research biologist and professor of conservation science, University of Cambridge.

Scottish wildcat: Adrian Davis, director, WildOutdoors; Rachel Williams, Highland Wildlife Park; Douglas Richardson, Highland Wildlife Park; Hannah and family; Dr David Hetherington, manager, Cairngorms Wildcat Project; Amy Cox, Highland Tiger Project; Dr Ruairidh Campbell, Wildlife Conservation Research Unit, University of Oxford.

Pine marten: John Picton, Speyside Wildlife guide; Warwick Lister-Kaye, Aigas Field Centre; David Bavin, pine marten project officer, Vincent Wildlife Trust.

Capercaillie: Andy Bateman, Fraoch Lodge, Boat of Garten; RSPB staff, Loch Garten.

Slavonian grebe: RSPB staff, Scotland office.

Streaked bombardier beetle: Dr Sarah Henshall, lead ecologist, Buglife; Matt Shardlow, chief executive, Buglife; Richard Jones, Royal Entomological Society.

Basking shark: Jackie and Graham Hall, Manx Basking Shark Watch; Natasha Phillips, marine biologist and research assistant, Manx Basking Shark Watch; Haley Dolton, research assistant, Manx Wildlife Trust; Professor David Sims, Marine Biological Association; Dr Peter Richardson, biodiversity programme manager, Marine Conservation Society; Dr Matthew Witt, University of Exeter; Dr Suzanne Henderson, Scottish Natural Heritage.

Montagu's harrier: Jim Scott, RSPB harrier protection scheme manager, East Anglia; Bob Image, harrier protection warden; Mark Thomas, investigations officer, RSPB.

Pool frog: John Baker, amphibian and reptile consultant; John Buckley, amphibian conservation officer, Amphibian and Reptile Conservation; Jim Foster, conservation director, Amphibian and Reptile Conservation.

Norfolk hawker dragonfly: Pam Taylor, former president, British Dragonfly Society; Claire Install, conservation officer, British Dragonfly Society.

Fen raft spider: Dr Helen Smith, president, British Arachnological Society.

Spiny seahorse: Neil Garrick-Maidment, executive director, the Seahorse Trust; Studland seahorse survey volunteers John and Eva; Tracey and Paul, Studland Watersports.

Shrill carder bee: Sinead Lynch, conservation officer, Bumblebee Conservation Trust; Kevin Dupé, reserves manager, Newport Wetlands; Michelle Appleby, Bumblebee Conservation Trust.

Silurian moth: Dave Grundy, wildlife consultant, dgcountryside; Martin Anthoney, Monmouthshire Butterfly and Moth Group; George Tordoff, Butterfly Conservation, Wales; Chris Manley; Chris Williams; 2013 Hatterrall Ridge Silurian survey group members; Tony Davis, Butterfly Conservation.

Wart-biter cricket: Peter Forrest, downlands warden, Kent Wildlife Trust; Greg Hitchcock, conservation officer, Kent Wildlife Trust; Dominic Price, director, Species Recovery Trust.

Black rat: Jim Lennon, Shiants Auk Ringing Group (SARG); David Steventon, SARG; Alister Clunas, SARG; Ian Buxton, SARG; Adam Nicolson, author of *Sea Room*; Tom Nicolson, owner of the Shiant Isles; Seumas Morrison, Sea Harris; Andi Dunkel, SkyXplorer; SARG members Bob Medland, Charlie Main, Ruth Walker, Alice Tribe, Karen Murray and Carole Davis; BBC Coast crew Miranda Krestovnikoff, Dan Davis, Brian Osborne and Phyllida O'Neil; Paul Walton, species and habitats policy officer, RSPB Scotland; David Wembridge, mammal surveys coordinator, People's Trust for Endangered Species.

Vendace: Dr Ian Winfield, Centre for Hydrology and Ecology; Janice Fletcher and Ben James, fish ecologists, Centre for Hydrology and Ecology; Dr Colin Bean, senior science and policy adviser, Scottish Natural Heritage; Andy Gowans, Environment Agency.

Black-winged stilt: RSPB staff, Pagham Harbour reserve; Medmerry RSPB stilt watch volunteers.

Additional species information: Dr Peter Evans, director, Sea Watch Foundation; Dr Russell Wynn, National Oceanography Centre; Paul Freestone, Cornwall Birding Association; Stephanie Morren, RSPB; Kevin Rylands, regional conservation adviser, RSPB South West.

Thanks also to many patient listeners, including colleagues at *The Herald*, Plymouth – Clare, Pete, Max, Paul A, Helen, Jayne, Edd, Neil, Nicky, Gillian, Hilary, Lisa, Keith, Tristan, Graham, Martin, Ian and Paul B among others – and to Dartmoor friends, including Stuart and Jackie, Jason and Susannah, Martin and Alice, John and Angela, Chris and Karole, Dom and Philippa, Josephine and Coll, Myles and Emma, Sarah and Chris, Dave and Judy, John and Colin, Ian, Jo and Chris, and Fiona. Also Laurence, Steve B, Steve O and Marcus, and to family, including Tess, Mark, Ben, Noreen, Paul and Lindsay, the Rees family, Elder family and Leach family, and with thoughts of Tim, grandparents, Maurice and Nick. Not forgetting Oakey, for getting me outdoors every day.

**Updates**

Several seasons have passed since some of my trips, and, while details were as accurate as possible at the time of writing, a few updates may be of interest. In 2014, no seahorses were found at the South Beach survey site at Studland, despite extensive searches. One transient juvenile was spotted off Middle Beach. The Seahorse Trust continues its campaign for the bay to become a Marine Conservation Zone. Meanwhile, Scotland's coastal waters off Oban, stretching from Loch Sunart to the Sound of Jura, were designated a Marine Protection Area to protect common skate. Fen raft spider populations expanded in translocation sites in the Norfolk and Suffolk Broads during 2014, while black rats in the Shiant Isles faced an uncertain future as funding was secured to remove them from the islands as part of a seabird recovery project. Two adult vendace were found in Bassenthwaite Lake in Cumbria during an autumn 2014 survey, following on from the single juvenile in 2013 – encouraging news given they had been considered extinct in this lake, which neighbours their Derwentwater stronghold. Only four pairs of Montagu's harriers were recorded in the UK in 2013; however, this improved in 2014, with twenty-three juveniles fledging from seven nests. There was less good news concerning golden orioles at Lakenheath RSPB reserve, with none recorded in 2014. Streaked bombardier beetles appeared to enjoy a good summer at their east London Docklands site – Buglife found a healthy congregation in September 2014, though the brownfield site remained earmarked for development.

# Further Reading

My thanks to the many authors, journalists and academics whose books, reports and papers provided invaluable background information. I won't attempt a comprehensive bibliography; however, a few texts worth mentioning include:


Barkham, P., *The Butterfly Isles* (Granta, 2011).
Benjamin, A. and McCallum, B., *A World Without Bees* (Guardian Books, 2009).
Brown, A. and Grice, P., *Birds in England* (Poyser, 2005).
Buczacki, S., *Fauna Britannica* (Hamlyn, 2002).
Chinery, M., *Collins Complete Guide to British Insects* (HarperCollins, 2009).
Clover, C., *The End of the Line* (Ebury, 2005).
Cocker, M. and Mabey, R., *Birds Britannica* (Chatto & Windus, 2005).
Dinerstein, E., *The Kingdom of Rarities* (Island Press, 2013).
Fuller, E., *The Great Auk* (Errol Fuller, 1999).
Gaston, K. J., *Rarity* (Chapman & Hall, 1994).
Green, R. and Riley, H., *Corncrakes* (Scottish Natural Heritage, 1999).
Hart-Davis, D., *Fauna Britannica* (Weidenfeld & Nicolson, 2002).
Holden, P. and Cleeves, T., *RSPB Handbook of British Birds* (Bloomsbury, 2014).
Inns, H., *Britain's Reptiles and Amphibians* (WILDguides, 2011).
Jepson, P. and Ladle, A., *Conservation* (Oneworld, 2010).
Lovegrove, R., *Silent Fields* (Oxford University Press, 2008).
Maclean, N. (ed.), *Silent Summer* (Cambridge University Press, 2010).
Mason, P. and Allsop, J., *The Golden Oriole* (Poyser, 2008).
Mynott, J., *Birdscapes: Birds in our imagination and experience* (Princeton University Press, 2012).
Nicolson, A., *Sea Room* (HarperCollins, 2002).
Oates, M., *Butterflies* (National Trust, 2011).
People's Trust for Endangered Species, *Britain's Mammals: a concise guide* (Whittet, 2010).
Salmon, M., *The Aurelian Legacy* (Harley, 1997).
Smith, M., *The British Amphibians and Reptiles* (HarperCollins, 1969).
Spicer, J., *Biodiversity* (Oneworld, 2006).
Sterry, P., *Collins Complete Guide to British Wildlife* (HarperCollins, 2008).
Tait, M., *Going, Going, Gone* (Think Books, 2006).
Taylor, M., *RSPB British Birds of Prey* (Christopher Helm, 2010).
Taylor, M., *Dragonflight* (Bloomsbury, 2013).
Thompson, K., *Do We Need Pandas?* (Green Books, 2010).
Titchmarsh, A., *British Isles: A Natural History* (BBC Books, 2004).
Titchmarsh, A., *The Nature of Britain* (BBC Books, 2007).

Useful reports, of which there have been many, include:

State of Nature 2013.
State of Britain's Larger Moths 2006/13, Butterfly Conservation/Rothamsted Research.
The State of the UK's Birds 2010/12/13.
The State of Britain's Mammals 2011, PTES.
Silent Seas, Marine Conservation Society, 2008.
Lost Life: England's Lost and Threatened Species, Natural England, 2010.
State of the Natural Environment 2008, Natural England.
British Seahorse Survey 2011, The Seahorse Trust.
Bechstein's Bat Survey 2007–2011, Bat Conservation Trust.
UK Conservation of Streaked Bombardier Beetle 2012, Eleanor Passingham, PTES/Buglife.
The Cairngorms Wildcat Project Final Report, Cairngorms National Park, 2012.
The Natural Choice: Securing the Value of Nature, HM Government, 2011.

# Useful Websites

Amphibian and Reptile Conservation: arc-trust.org
ARKive: arkive.org
BBC: bbc.co.uk
BirdGuides: birdguides.com
British Dragonfly Society: british-dragonflies.org.uk
British Trust for Ornithology: bto.org
Buglife: buglife.org.uk
Bumblebee Conservation Trust: bumblebeeconservation.org
Butterfly Conservation: butterfly-conservation.org
DEFRA: defra.gov.uk
Froglife: froglife.org
International Union for Conservation of Nature: iucn.org
Manx Basking Shark Watch: manxbaskingsharkwatch.com
Marine Conservation Society: mcsuk.org
Natural England: naturalengland.org.uk
People's Trust for Endangered Species: ptes.org
Rare Bird Alert: rarebirdalert.co.uk
Species Recovery Trust: speciesrecoverytrust.org.uk
The Mammal Society: mammal.org.uk
The RSPB: rspb.org.uk
The Shark Trust: sharktrust.co.uk
The Wildlife Trusts: wildlifetrusts.org
Vincent Wildlife Trust: vwt.org.uk
Wikipedia: en.wikipedia.org
Wild About Britain: wildaboutbritain.co.uk
Wildlife Extra: wildlifeextra.com
Zoological Society of London: zsl.org

# Index